Daniel Karthe, Sergey Chalov,
Nikolay Kasimov, Martin Kappas (eds.)

Water and Environment in the Selenga-Baikal Basin:

International Research Cooperation
for an Ecoregion of Global Relevance

EARTH VIEW - GEOGRAPHY AND GEOINFORMATION

Edited by Prof. Dr. Martin Kappas

ISSN 1614-4716

Daniel Karthe, Sergey Chalov,
Nikolay Kasimov, Martin Kappas (eds.)

WATER AND ENVIRONMENT IN THE SELENGA-BAIKAL BASIN:

INTERNATIONAL RESEARCH COOPERATION FOR AN ECOREGION OF GLOBAL RELEVANCE

ibidem-Verlag
Stuttgart

Bibliographic information published by the Deutsche Nationalbibliothek
Die Deutsche Nationalbibliothek lists this publication in the Deutsche Nationalbibliografie;
detailed bibliographic data are available in the Internet at http://dnb.d-nb.de.

Bibliografische Information der Deutschen Nationalbibliothek
Die Deutsche Nationalbibliothek verzeichnet diese Publikation in der Deutschen
Nationalbibliografie; detaillierte bibliografische Daten sind im Internet über http://dnb.d-nb.de
abrufbar.

ISSN: 1614-4716

ISBN-13: 978-3-8382-0863-3

© *ibidem*-Verlag / *ibidem* Press

Stuttgart, Germany 2016

Printed in the United States of America

Contents

VI

Water and Environment in the Selenga-Baikal Basin: International Research Cooperation for an Ecoregion of Global Relevance

Daniel Karthe[1,2], Sergey Chalov[3], Nikolay Kasimov[3], Martin Kappas[2]

[1] Department Aquatic Ecosystem Analysis and Management, Helmholtz Centre for Environmental Research, Magdeburg, Germany

[2] Institute of Geography, Georg-August University, Göttingen, Germany

[3] Faculty of Geography, Lomonosov Moscow State University, Russia

The Selenga Baikal Basin: An Ecoregion of Global Importance

As the deepest and oldest lake in the world, Lake Baikal features a unique ecosystem which was declared a world natural heritage site by the United Nations in 1996. The lake's most important tributary is the Selenga River, which contributes about 50% of the influx into Lake Baikal and has a consierable part of its runoff generated in the Mongolian part of its basin (Chalov et al. 2015; Karthe et al. 2013). Together with the Angara and Yenisey rivers it forms the longest river network in Eurasia and has been widely recognized as a significant driver for the state of Lake Baikal.

Large parts of the Selenga River Basin, and in particular the upstream subcatchments of the Selenga and its tributaries, are sparsely settled and still in a relatively natural state. These headwater regions are often mountainous, forested and responsible for a large part of the runoff generation (Minderlein & Menzel 2015). Rivers have a largely natural hydrological regime without notable abstractions or hydromorphological modifications. Further downstream, meandering river courses flowing through wide valleys covered mostly by steppe are characteristic. Parts of these grasslands have been converted to cropland, which is often irrigated due to the low precipitation (below 500 mm for most of the river basin) and high evapotranspiration (Karthe et al. 2014; Menzel et al. 2011). Moreover, large herds of livestock (sheep, goats, cattle, horses, camels) are reared in the Mongolian part of the Selenga River Basin. Agricultural water demand had been increasing over the past decade and is expected to rise further due to a warming trend that is about twice the global average (Malsy et al. 2015; Törnquist et al. 2014). The three largest cities of Mongolia (Ulaanbaatar, Erdenet and Darkhan) as well as Ulan Ude, the capital of the Republic of Buryatia in Russia, are located on the Tuul, Orkhon, Kharaa and Selenga Rivers, respectively. Because of the associated concentrations of population and industry, these cities are major water users and waste water producers (Gardemann et al. 2012; Karthe et al. 2015). In addition, various mining activities are concentrated in the basin, including the exploitation of coal, gold, copper, molybdenum and wolfram (Sandmann 2012; Thorslund et al. 2012). As a consequence, problematic concentrations of nutrients, heavy metals and other toxic substances have recently been detected in surface and groundwater resources (Avlyush 2011; Hofmann et al. 2015; Nadmitov et al. 2014; Pfeiffer et al. 2015). At the same time, the transboundary location of the Selenga river system means that gathering consistent information on water availability and quality across state borders constitutes a challenge for regional water management Given the current development of the irrigation and mining sectors as well as plans for the construction of dams and reservoirs in the upper parts of the Selenga River, these problems could become exacerbated in the future (Chalov et al. 2015).

North of Ulan Ude, the Selenga River branches into a wide delta, the largest freshwater inland delta in the world. The associated wetland constitutes a unique ecosystem (Гармаев & Христофоров 2010) and acts as the final geobiochemical barrier before the Selenga discharges into Lake Baikal. It has a great impact on pollution delivery to Lake Baikal, storing up to 60-70 % of the sediment load of the Selenga River (Тулохонов & Плюснин 2008).

Bringing Together Selenga Baikal Research: An Interdisciplinary Collaboration of International Experts

Given its large scale (see figure 1), global relevance and the numerous water-related challenges, it is little surprise that the water resources and aquatic eco-systems of the Selenga river and Lake Baikal basin were and still are the sub-ject of various national and international research projects. Involving a multi-tude of regional and foreign experts, the projects have not only dealt with a wide range of geoscientific, biological, economic and socio-political topics, but also focused on different parts of the large Selenga River Basin.

Figure 1: The size of Lake Baikal, as illustrated by one of the participants of the Bringing Together Selenga-Baikal Research Conference (Prof. Dr. Christian Opp, Marburg University, Germany)

During the past 5 years, several research projects which had started independently from each other began exchanging data and coordinating their field activities.This lead to the development of a collaboration between Lomonosov Moscow State University (Russia), Stockholm University (Sweden), the Helmholtz Centre for Environmental Research (Germany) and École polytechnique fédérale de Lausanne (EPFL, Switzerland) who in 2012 co-organized an international and interdisciplinary workshop, the "Bringing Together Selenga Baikal Research Conference". The first such conference was held at the international campus of Lomonosov Moscow State University in Geneva, Switzerland, and attended by experts from France, Germany, Mongolia, Russia, Sweden, Switzerland, and the USA.

In 2013, representatives of several groups conducted a joint expedition through the Selenga River Basin which was followed by the "Baikal – Strategic Global Resource of the 21st Century" conference in Ulan Ude, thereby continuing the discussion and exchange of ideas that had been initiated the year before.

In October 2014, the Helmholtz Centre for Environmental Research in Leipzig, Germany hosted the second "Bringing Together Selenga Baikal Research Conference". The conference was attended by more than 40 international experts, some of them with several decades of research experience in the study region. The papers in this book, which are based on the conference, provide an overview about the objectives and key findings of past and current research activities in the Selenga-Baikal Basin. In this way, they summarize the present state of the art and but also indicate future research needs.

Photos: Participants of the Bringing Together Selenga-Baikal Research Conference on 31 December 2012 in Geneva, Switzerland at Moscow State University International Campus, and 2 October 2014 at the Helmholtz Centre for Environmental Research, Leipzig, Germany

Key Results of the 2014 Bringing Together Selenga Baikal Research Conference

A first group of papers deals with the **availability of surface and groundwater resources and the role of rising abstractions, climate and land use change** in this context. *Malsy & Flörke* utilized the WaterGAP3 model and scenarios for climate change and socioeconomic development to forecast future water availability and abstraction trends. *Kappas, Renchin, Munkhbayar, Vova and Degener* discuss the feasibility of long-time satellite data series for the detection of land use change. Such information is valuable since land cover changes are very pronounced in some parts of the Selenga-Baikal Basin, and are expected to have significant impacts on the regional hydrology. *Renchin, Kappas, Munkhbayar, Vova and Degener* present an example of possible drivers of land degradation in a Mongolian province. *Dandar et al.* present first results of a study on groundwater recharge in the Mongolian national capital region. Their study shows that (a) groundwater is a very limited resource that can easily be overexploited in densely settled areas such as Ulaanbaatar, and that (b) quantitative assessments are restrained by poor data availability.

A second group of papers deals with **environmental pollution impacts of anthropogenic activities**. *Batbayar et al.* analysed the water quality pattern in surface water bodies in Northern Mongolia, identifying pollution gradients that follow urban and mining impacts. *Shinkareva et al.* discuss the development of heavy metal fluxes from mining areas and major urban centers to Lake Baikal. *Chalov & Romanchenko* considered the impacts of environmental changes in the river basin on sediment and pollutant transport. The three papers show that anthropogenic activities in the Selenga Baikal Basin have manifold impacts on water quality, including increased sediment influx, rising nutrient loads and contamination by heavy metals. *Koshaleva et al.* investigated the levels of soil contamination in six urban and mining areas along the Selenga and its tributaries. Two of the areas discussed here are described in more detail in the case studies presented by two additional papers. *Timofeev* assessed the effects of mining on soil contamination in the city of Zakamensk, where pollution levels in two thirds of the urban area were found to be significant. *Sorokina* created a geochemical map of Ulaanbaatar which localizes major pollution sources and distinguishes between zones characterized by different levels of environmental pollution.

In addition to the papers dealing with sediment transport primarily from a water quality perspective, two more contributions focus on **fluvial transport dynamics and morphology**. *Promakhova & Alexeevsky* analyse the results of field measurements of sediment loads in 2013-2014 done by Lomonosov MSU team together with Stockholm University and describe the sediment transport regime of the Selenga under different runoff conditions, thereby looking at the entire length of the river system. One particularly relevant outcome of this study is that noted longitudinal inequalities in sediment balances could increase after the planned construction of dams along the Selenga. *Le Dantec* et al. investigated the Kukuy Canyon (Lake Baikal) by shallow bathymetry, including a discussion of sediment transport and deposition pattern from the nearby mouth of the Selenga.

The **state of aquatic and terrestrial ecosystems** and relevance of specific stressors is the focus of a fourth group of articles. *Enkh-Amgalan et al.* provide an overview of the natural environment and anthropogenic stressors in the Mongolian part of the Selenga River Basin. *Gunin & Bazha* analysed the modification and degradation of natural vegetation in the entire Selenga-Baikal Basin and compared it to the developments in neighboring regions of Central Asia. *Luckenbach et al.* show that climate change is likely to cause regional shifts and changes in the composition of the amphipod fauna of Lake Baikal. *Shimaraeva et al.* compared the development of Baikal phytoplankton in a reference area without waste water influence and the region impacted by treated waste water from the pulp and paper factory in Baikalsk (which was shut down in 2013).

Water management is the central theme of a fifth section consisting of four papers. *Garmaev et al.* describe the history and categorization of flood events for the Russian part of the Selenga River Basin, including the relevance of hydrological research for better preparedness. *Karthe et al.* summarize the scientific basis for the conceptualization of an Integrated Water Resources Management (IWRM) in the Mongolian Kharaa River Basin. In the same context, *Heldt et al.* discuss to what degree the European Water Framework Directive (EU-WFD) could serve as a model for river basin management planning in Mongolia. *Westphal et al.* assessed potential realizations of constructed wetlands for waste water treatment in Mongolian cities.

The final section introduces **innovative monitoring techniques** that may be relevant for future water-related investigations in the Selenga-Baikal Basin. *Akhtman et al.* describe the development and deployment of a novel multispectral and hyperspectral remote sensing platform that is carried by ultralight aircraft. *Siegfried et al.* outline advantages of biosensors for water quality monitoring and take a specific look at the applicability of the recently developed ARSOlux system for screening and monitoring of drinking water in potentially arsenic-affected areas of Mongolia. The necessity of investigating drinking, ground and surface water hygiene in the Selenga River Basin is discussed by *Karthe*, including perspectives for the application for advanced microbiological techniques. Finally, *Gerhardt* introduces a freshwater biomonitor that can be used to detect water quality deteriorations by the responses of different indicator species, including pollution-sensitive gammarids which are naturally present in the Selenga-Baikal system.

The papers presented in this book are manuscripts present the views of the authors and have not been edited stylistically or content-wise. They range from consolidated findings to preliminary results of ongoing studies and future perspectives for research. Despite these differences, the collection of papers in this book constitutes a unique documentation of the current state of scientific knowledge on water issues in the Selenga-Baikal Basin – in particular because some of the findings presented here have previously not been published in English. Finally, the motivation of scientists representing several countries and disciplines to participate in the Bringing Together Selenga Baikal Research conference and to contribute a full paper to these proceedings demonstrate that there is a strong determination in the scientific community to share results and cooperate in order to come to a better understanding of water-related challenges in the Selenga-Baikal Basin.

References

Akhtman, Y.; Constantin, D.; Rehak, M.; Nouchi, V.; Tarasov, M.; Shinkareva, G.; Chalov, S. & Lemmin, U.: Leman-Baikal: Remote Sensing of Lakes Using an Ultralight Plane. *This volume*, pp. 323-333.

Batbayar, G.; Karthe, D.; von Tümpling, W.; Pfeiffer, M. & Kappas, M. (2015): Influence of urban settlement and mining activities on surface water quality in northern Mongolia. *This volume*, pp. 73-86.

Chalov S.R.; Jarsjö, J., Kasimov, N.; Romanchenko, A.; Pietron, J.; Thorslund, J. & Belozerova, E. (2015): Spatio-temporal variation of sediment transport in the Selenga River Basin, Mongolia and Russia. Environmental Earth Sciences 73(2):663-680.

Chalov, S. & Romanchenko, A. (2015): Linking Catchments to Rivers: Flood-driven Sediment and Contaminant Loads from catchment and in-channel sources in the Selenga River. *This volume*, pp. 101-118.

Dandar, E.; Ramirez, J.C.; Nemer, B. (2015): Evaluation of groundwater resources in the upper Tuul River basin, Mongolia. *This volume*, pp. 55-69.

Enkh-Amgalan, S.; Dorjgotov, D.; Oyungerel, J.; Enkh-Taivan, D. & Batkhishig, O. (2015): Geo-ecological Issues in the Selenga River Catchment. *This volume*, pp. 193-205.

Gardemann, E. & Stadelbauer, J. (2012): Städtesystem und regionale Entwicklung in der Mongolei: Zwischen Persistenz und Transformation. Geographische Rundschau 64(12):34-41. Publication in German. [Gardemann, E. & Stadelbauer, J. (2012): Urban system and regional development in Mongolia: between persistence and transformation. Geographische Rundschau 64(12):34-41].

Garmaev, E.; Borisova T.; Ayurzhanayev, A.; Tsydypov, B. (2015): Floods in the Selenga River basin: research experience. *This volume*, pp. 255-264.

Гармаев, Е.Ж. & Христофоров, А.В. (2010): Водные ресурсы рек бассейна озера Байкал: основы их использования и охраны. Новосибирск: Академическое издательство «ГЕО». Publication in Russian. [Garmaev, E.Zh. & Khristovorov, A.V. (2010): Water Resources of the Rivers of the Lake Baikal Basin: Basics of Their Use and Protection. Novosibirsk: Academic Press "Geo"]

Gerhardt, A.: The Multispecies Freshwater Biomonitor: Applications in Ecotoxicology and Water Quality Biomonitoring. *This volume,* pp. 347-354.

Gunin, P.& Bazha, S. (2015): Interaction of Ecosystems of the Selenga Basin and Environmental Risks in Central Asia. *This volume,* pp. 207-218.

Heldt, S.; Karthe, D.; Feld, C. (2015): The EU-WFD as an Implementation Tool for IWRM in non-European countries – Case Study: Mongolia. *This volume,* pp. 281-299.

Hofmann, J; Watson, V. & Scharaw, B. (2015): Groundwater quality under stress: contaminants in the Kharaa River basin (Mongolia). Environmental Earth Sciences 73(2):629-648.

Hofmann, J.; Venohr, M.; Behrendt, H. & Opitz, D. (2010): Integrated Water Resources Management in Central Asia: Nutrient and heavy metal emissions and their relevance for the Kharaa River Basin, Mongolia. Water Science and Technology 62(2)353-363.

Kappas, M., Renchin, T., Munkhbayar, S., Vova, O., Degener, J. (2015): Review of Long-term Satellite Data Series on Mongolia for the Study of Land Cover and Land Use. *This volume,* pp. 27-35.

Karthe, D.; Chalov, S.; Malsy, M.; Menzel, L.; Theuring, P.; Hartwig, M.; Schweitzer, C.; Hofmann, J.; Priess, J.; Shinkareva, G. & Kasimov, N. (2014): Integrating Multi-Scale Data for the Assessment of Water Availability and Quality in the Kharaa - Orkhon - Selenga River System. Geography, Environment, Sustainability 7(3): 65-86.

Karthe, D.; Chalov, S.; Theuring, P. & Belozerova, E. (2013): Integration of Meso- and Macroscale Approaches for Water Resources Monitoring and Management in the Baikal-Selenga-Basin. In: Chifflard, P.; Cyffka, B.; Karthe, D. & Wetzel, K.-F. (Eds.) (2013): Beiträge zum 44. Jahrestreffen des Arbeitskreises Hydrologie, pp. 90-94. Augsburg: Geographica Augustana

Karthe, D. & Heldt, S. (2015): Challenges for Science-Based IWRM Implementation in Mongolia: Experiences from the Kharaa River Basin. *This volume,* pp. 265-280.

Kosheleva, N.E.; Kasimov, N.S.; Gunin, P.D.; Bazha, S.N.; Sandag, E.-A.; Sorokina, O.; Timofeev, I.; Alexeenko, A. & Kisselyeva, T. (2015): Hot Spot Assessment: Cities of the Selenga River Basin. *This volume,* pp. 119-136.

Le Dantec, N.; Babonneau, N.; Franzetti, M.; Delacourt, C.; Akhtman, Y.; Ayurzhanaev, A. & Le Roy, P. (2015): Morphological analysis of the upper reaches of the Kukuy Canyon derived from shallow bathymetry. *This volume,* pp. 179-190.

Luckenbach, T.; Bedulina, D. & Timofeyev, M. (2015): Is the Endemic Fauna of Lake Baikal Affected by Global Change? *This volume,* pp. 219-235.

Malsy, M; aus der Beek, T. & Flörke, M (2015): Evaluation of large-scale precipitation data sets for water resources modelling in Central Asia. Environmental Earth Sciences 72(2):787-799.

Malsy, M. & Flörke, M. (2015): Large-scale modelling of water resources in the Selenga River. *This volume,* pp. 17-26.

Menzel, L.; Hofmann, J. & Ibisch, R. (2011): Untersuchung von Wasser- und Stoffflüssen als Grundlage für ein Integriertes Wasserressourcen – Management im Kharaa-Einzugsgebiet (Mongolei). Hydrologie und Wasserbewirtschaftung 55(2):88-103. Publication in German. [Menzel, L.; Hofmann, J. & Ibisch, R. (2011): Investigation of water and matter fluxes as the basis for an Integrated Water Resources Management in the Kharaa River Basin (Mongolia). Hydrologie und Wasserbewirtschaftung 55(2):88-103.]

Minderlein, S. & Menzel, L. (2015): Evapotranspiration and energy balance dynamics of a semi-arid mountainous steppe and shrubland site in northern Mongolia. Environmental Earth Sciences 73(2): 593-609.

Nadmitov, B.; Hong, S. Kang, S.I.; Chu, J.M.; Gomboev, B.; Janchivdorj, L.; Lee, C.H, & Khim, J.S. (2014): Large-scale monitoring and assessment of metal contamination in surface water of the Selenga River Basin (2007–2009). Environmental Science and Pollution Research International 22(4):2856-2867.

Pfeiffer, M.; Batbayar, G.; Hofmann, J.; Siegfried, K.; Karthe, D. & Hahn-Tomer, S. (2015): Investigating arsenic (As) occurrence and sources in ground, surface, waste and drinking water in northern Mongolia. Environmental Earth Sciences 73(2):649-662.

Opp, Ch. (1994): Naturphänomene und Probleme des Natur- und Umweltschutzes am Baikalsee. Petermanns Geographische Mitteilungen 138

(4):219-234. Publication in German. [Opp, Ch. (1994): Natural phenono-mena and problems of nature and environmental protection. Petermanns Geographical Notes 138 (4):219-234.]

Promakhova, E. & Alexeevsky, N. (2015): Source to Sink: Water and Sediment Transport in the Selenga-Baikal Catchment. *This volume*, pp. 167-178.

Renchin, T., Kappas, M., Munkhbayar, S., Vova, O., Degener, J. (2015): Drivers of land degradation in Umnugobi province. This volume, pp. 37-53.

Sandmann, R. (2012): Gier nach Bodenschätzen und Folgen für die Mongolei. Geographische Rundschau 64(12):26-33. Publication in German. [Sandmann, R. (2012): Greed for raw materials and its consequences for Mongolia. Geographische Rundschau 64(12):26-33].

Shimaraeva, S.V.; Izmestyeva, L.R.; Krashchuk, L.S.; Pislegina, H.V.; Silow, E.A.: The influence of BPPC on Baikal plankton – comparative study of phytoplankton in the point of influence of BPPC purified waste waters and in the reference clean point in 2005-2006 years. *This volume*, pp. 237-251.

Shinkareva, G.L.; Kasimov, N.S. & Lychagin, M.Y. (2015): Heavy Metal Fluxes in the Rivers of the Selenga Basin. *This volume*, pp. 87-100.

Siegfried, K.; Koelsch, A.; Osterwalder, E. & Hahn-Tomer, S. (2015): Advantages of Biosensor Water Quality Monitoring. *This volume*, pp. 335-346.

Sorokina, Olga (2015): Environmental-Geochemical Map of Ulaanbaatar City: Methodology of Compiling and Perspectives of Applying. This volume, pp. 153-164.

Timofeev, I. (2015): Geochemical Transformation of Soils Caused by Non-Ferric Ore Mining in the Selenga River Basin (Case Study of Zakamensk). *This volume*, pp. 137-151.

Thorslund, J.; Jarsjö, J.; Chalov, S. & Belozerova, E. (2012): Gold mining impact on riverine heavy metal transport in a sparsely monitored region: the upper Lake Baikal Basin case. Journal of Environmental Monitoring 14(10): 2780–2792.

Törnqvist, R.; Jarsjö, J.; Pietron, J.; Bring, A.; Rogberg, P.; Asokan, S.M. & Destouni, G. (2015): Evolution of the hydro-climate system in the Lake Baikal basin.

Тулохонов, А.К. & Плюснин, А.М. (Eds.) (2008): Дельта реки Селенги – естественный биофильтр и индикатор состояния озера Байкал. Отв. Новосибирск: изд-во СО РАН. Publication in Russian. [Tulokhonov, A.K. & Plyusnin, A.M. (Eds.) (2008): Selenga River Delta as a Natural Biofilter and Indicator of the State of Lake Baikal. Novosibirsk: SO RAN Publications.]

Westphal, K.; Sullivan, C.; Gregersen, P. & Karthe, D. (2015): Potential and feasibility of willow vegetation filters in Mongolia. *This volume*, pp. 300-320.

I. Availability of surface and groundwater resources and the role of rising abstractions, climate and land use change

Large-scale modelling of water resources in the Selenga River Basin

Marcus Malsy and Martina Flörke

Center for Environmental Systems Research, University of Kassel, Germany

Corresponding author: malsy@usf.uni-kassel.de

Abstract

The Selenga River Basin contributes more than 60% to Lake Baikal inflows. Beside changes in the hydro-climatic system water pollution is a rising issue. At this, mostly agricultural and industrial activities contribute to surface water pollution. Nevertheless, data and information about abstractions for sectoral water use purposes are very scarce. In this study, spatial-explicit quantification of consumptive water use for agricultural, industrial, and domestic uses in the Selenga River Basin is calculated. These abstractions were computed with the global water resources model WaterGAP3 for the base year 2005 and the scenario year 2055 under the shared socio-economic pathway SSP2 and the representative concentration pathway RCP 6.0. Climate simulations from five General Circulation Models (GCM), namely IPSL-CM5A-LR, MIROC-ESM, NorESM1-M, HadGEM2-ES and GFDL-ESM2M, were used. The results show an increase in all five sectors by 2055 with an extreme trend to higher water abstractions in the manufacturing sector, which is triggered by a strong economic development in the global scenario. In total the calculated water abstractions increase from 163.54 Mio m^3a^{-1} in 2005 to 298.28 Mio m^3a^{-1} in 2055.

Introduction

Recent studies for Mongolia focussed on aspects of current and future changes in water quantity (e.g. Batimaa 2006, Menzel et al. 2008, Malsy et al. 2012, Törnqvist et al. 2014) and/or water quality (e.g. Thorslund et al. 2012, Hofmann et al. 2013, Karthe et al. 2014), but mostly did not pay attention to current and future water uses for anthropogenic purposes. Batsukh et al. (2008) and Malsy et al. (2013) showed that water use abstractions play an important role in northern Mongolia and are expected to increase in future due to urbanisation, increasing population, and rising industrial activities. These water abstractions affect both water quality (e.g. water used for ore washing in mining) as well as water quantity. At this, water withdrawals are highest in the mining sector followed by households and livestock. In this study, the integrated large-scale hydrological, water use, and water quality model WaterGAP3 (Alcamo et al. 2003, Döll et al. 2003, aus der Beek et al. 2010, Flörke et al. 2012, Flörke et al. 2013) is used to simulate current (2005) and future (2055) sectoral water consumptive uses for livestock, irrigation, manufacturing, thermal electricity production, as well as domestic and small businesses purposes to quantify current and future impacts on water resources in the Selenga River Basin.

Model and Data

The WaterGAP3 model (Verzano 2009) is a further development of WaterGAP2 (Alcamo et al. 2003) and is based on a five by five arc minutes grid (~6x9 km) with daily internal time steps. For each grid cell, a water balance is calculated under consideration of climate time series and physical geographic data, e.g., land cover and soil texture. Furthermore, sectoral water uses are computed for agricultural (irrigation, and livestock), manufacturing industry, thermal electricity production, and domestic and small business purposes (aus der Beek et al. 2010, Flörke et al. 2012, Flörke et al. 2013). At first, water abstractions for the sectors livestock, irrigation, manufacturing, thermal electricity production as well as domestic and small business were computed using the socio-economic and energy-related drivers following the Shared Socio-economic Pathway (SSP) 2 (van Vuuren et al. 2011, O'Neill et al. 2014) and fed into the hydrological model. Afterwards, hydrological fields, such as evapotranspiration, were simulated for current and future climate conditions. For this purpose, data from the general circulation models IPSL-CM5A-LR, MIROC-ESM, HadGEM2-ES, GFDL-ESM2M, and NorESM1-M, driven by the Representative Concentration Pathway (RCP) 6.0, were used.

Results and Discussion

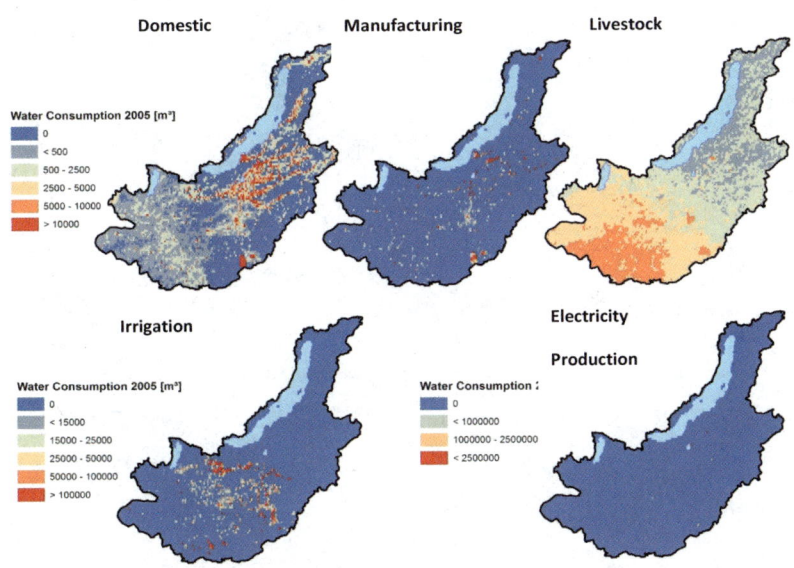

Figure 1: Water consumption in the Selenga-Baikal River basin for the base year 2005

The water use sectors (cf. Fig. 1) show a diverse picture for the Selenga-Baikal River basin. The domestic sector has the highest abstractions southeast of Lake Baikal, while the south-western part shows small, but area-covering abstractions. Manufacturing abstractions focus on cities and are therefore less densely spread, but feature generally higher abstractions. Water use for livestock is distributed throughout the entire river basin with higher abstractions especially in the Mongolian part of the river basin. However, the amount of livestock water consumption is generally in lower categories. Water abstractions for irrigation are clustered in the central part with high abstractions in the Dzhida, and Kharaa sub-basins. This is even more pronounced for thermal electricity production, which occurs only in a few places but features very high local water consumption. For a detailed description of the Selenga River basin and its sub-basins see Karthe et al. 2014.

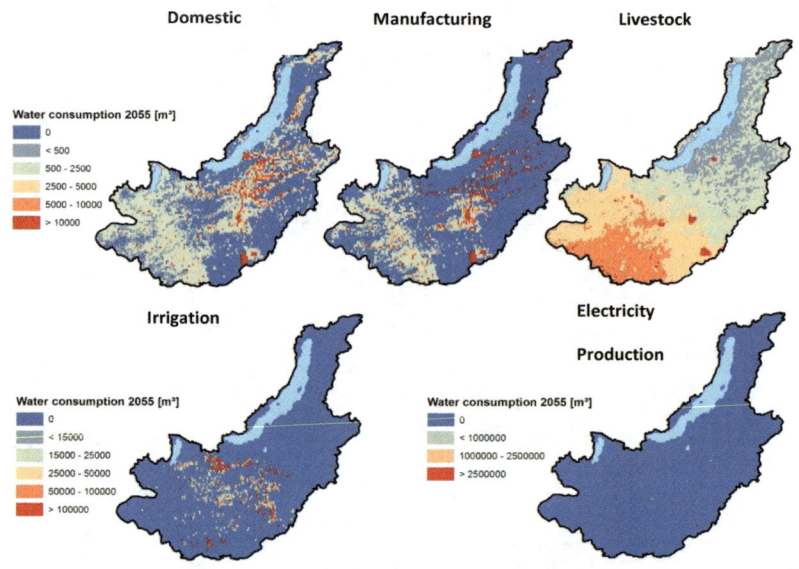

Figure 2: Water consumption in the Selenga-Selenga Baikal River basin for the scenario year 2055

Future water consumption (cf. Fig. 2) shows an increase in the south western part, particularly for manufacturing but also for the domestic sector. Spatial patterns remain the same for livestock, irrigation, and electricity production sectors. Manufacturing increases in the sub-basins of Uda and Khilok River. This is also mirrored in total and relative sectoral contributions to total water consumption (cf. Table 1). Hereby, manufacturing rises from 9.21 Mio. m³a⁻¹ to 121.27 Mio. m³ a⁻¹ in 2055 with a relative sectoral contribution rising from 5.6% to 40.7%. This increase in manufacturing water consumption is mainly triggered by a rapid increase of the Gross Value Added (GVA) till 2055 in SSP2

(O'Neill et al. 2014), which is a main driver of the manufacturing sector (cf. Flörke et al 2013). According to this high rise in the manufacturing sector, all other sectoral shares decrease in 2055, but nevertheless also show an increase of water consumption in absolute terms (cf. Table 1).

Considering other water use studies of Batsukh et al. 2008, and Malsy et al. 2013, it is difficult to make a comparison as both estimated the water uses for entire Mongolia, while in this study just the Selenga River Basin was examined. To our knowledge, there are no recent estimations of water consumption in the Russian part of the Selenga River Basin. Furthermore, this study focusses on water consumption as opposed to water withdrawals, which are the basis of Batsukh et al. 2008. However, as the Selenga River Basin is the main river basin in Mongolia and features most of the available water resources, a relative comparison to Batsukh et al 2008 is possible. They estimated for 2005/2006 the highest sectoral water abstractions for mining industry, followed by hydro-power plants, drinking water supply, and livestock with an overall water use of 433.78 Mio. m³. Malsy et al. 2013 estimated 800 Mio m³ water withdrawals for 2005, which is mainly driven by much larger abstractions for electricity production and irrigation compared to Batsukh et al. 2008.

Year	Domestic	Manufacturing	Irrigation	Livestock	Electricity Production
2005	39.95	9.21	73.49	27.97	12.92
2055	43.32	121.27	84.38	29.38	19.93
Year	Domestic	Manufacturing	Irrigation	Livestock	Electricity Production
2005	24.4	5.6	44.9	17.1	7.9
2055	14.5	40.7	28.3	9.8	6.7

Table 1: Comparison of sectoral water consumption [Mio. m³*a⁻¹] (top) and [%] (bottom) (base year 2005 - scenario 2055)

The Korean Environment Institute (KEI 2008) estimated 919 Mio m³ of water withdrawals in the Selenge River Basin for the year 2004 with an abstraction from surface waters of 70%. After utilisation, 660.0 Mio m³ where returned to the river network, which leads to a water consumption of 258.4 Mio m³. In terms of the sectoral shares of water consumption, KEI (2008) reports the highest water consumption in the agricultural sector with 123.7 Mio m³, followed by in industry with 46 Mio m³ and by 30.9 Mio m³ for housing and communal services. 57.8 Mio m³ water consumption in the "other" sector is unfortunately not explained in more detail. Priess et al. 2011 reported an increasing competition for water between the water use sectors in the Kharaa River. Furthermore, the agricultural sector, and hence the use of scarce water resources, is expected to grow significantly, "motivated by subsidised water fees, irrigation equipment and cheap loans" (Hantulga 2009 cited in Priess et al 2011). Generally, high losses from leakages of around 50% in piped water systems can be observed (Scharaw & Westerhoff 2011), which is partly mirrored by large per-capita values of daily domestic water use in urbanized areas compared to much lower values in ger districts and local herders (Batsukh et al. 2008).

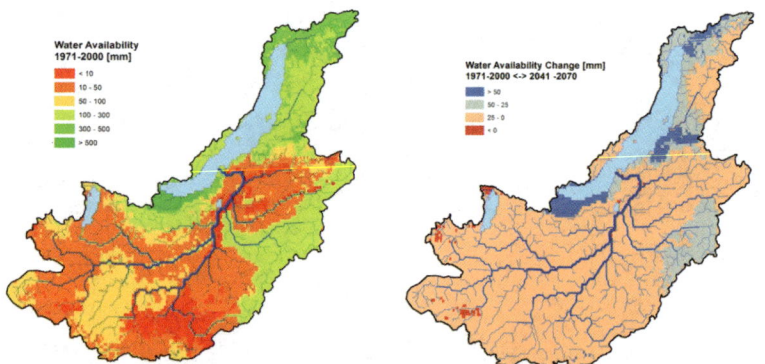

Figure 3: Modelled water availability for baseline time period 1971 - 2000 (left) and change to scenario period 2041-2070 (respectively GCM mean)

Current mean annual water availability (WA) (cf. Fig. 3) shows large parts with low water resources, especially in the southern Tuul catchment. Furthermore, also the Orkhon, Ider, Chulut, Delgermöron, Eg, Uda, and Khilok sub-basins feature wide areas with less than 50 mm mean annual water availability. The highest water availability above 500 mm a⁻¹ can be found in the Upper Angara river basin and at the south-western shore of Lake Baikal with a maximum water availability of 890 mm a⁻¹. Spatially, these baseline simulations (GCM mean, 1971-2000) show a high accordance compared with simulations conducted with the reanalysis WATCH forcing data (see Karthe et al. 2014). However, recent studies reported an overestimation of GCM baselines compared to reanalysis data (e.g. Malsy et al. 2012, Törnqvist et al. 2014). This aspect was beyond the scope of this study but might affect the total amount of available water resources leading to even lower available water resources. Future conditions depict an increasing trend till 2055 for the Chikol, and Khilok headwater, the upper Angara, and the Turka River. River basins with decreasing water availability can be found, e.g. Chuluut, Ider, Delgermöron and Eg river basin, but are overall less dense than river basins with increasing or stagnating WA. The mean annual WA for the entire Selenga River Basin is 109 mm in 2005 and rises to 124 mm in 2055, also the maximum value increases to 975 mm. According to Törnqvist et al. (2014), the temperature has been risen in the Selenga River Basin between 1938 and 2009 twice as much than the global average with decreasing inter-annual runoff variability, which points to permafrost thawing and therefore higher soil storage volumes.

Conclusions and Outlook

Generally, the water use projections show an increase of consumptive water uses by 2055 of 82.4% compared to 2005, particularly in the manufacturing sector. Furthermore, the manufacturing and domestic sectors spread spatially in the south western part of the Selenga River Basin and in the sub-basins around Ulan-Ude. Compared with Batsukh et al. 2008 and Malsy et al. 2013, mining plays an important role as water user but could not be included in this study as no data was available for the Russian part of the basin. At this, also large-scale impacts on water quality by mining, e.g., heavy metals and total dissolved solids, should be examined spatially explicit. Water availability shows an increasing trend in the northern part of the basin around Lake Baikal. A decreasing trend could be derived in the most western parts of the basin. As Permafrost melting is not included in this modelling study effects occurring due to permafrost thawing and their impacts on the water cycle could not be examined in this study, though they play a major role due to increasing soil storage water volume (cf. Menzel et al. 2008, Törnqvist et al. 2014). As Törnqvist et al. (2014) showed for water availability, future impacts on water uses should also be studied for all SSP and RCP combinations to get the full range of future projections.

References

Alcamo J., Döll P., Henrichs T., Kaspar F., Lehner B., Rösch T. and Siebert S. (2003): Development and Testing of the WaterGAP 2 Global Model of Water Use and Availability. Hydrological Science 48 (3): 317-337.

Aus der Beek T., Flörke M., Lapola D.M., Schaldach R., Voß F. and Teichert E. (2010): Modelling historical and current irrigation water demand on the continental scale: Europe. Adv. Geosci. 27: 79-85.

Batima P. (2006): Climate Change Vulnerability and Adaptation in the Livestock Sector of Mongolia. A Final Report Submitted to Assessments of Impacts and Adaptations to Climate Change (AIACC), Project No. AS 06, Washington, 105pp.

Batsukh N., Dorjsuren D. and Batsaikan G. (2008): The water resources, use and conservation in Mongolia. National Water Committee, Ulaanbaatar. Mongolia.

Döll, P., Kaspar, F., Lehner, B. (2003): A global hydrological model for deriving water availability indicators: model tuning and validation. Journal of Hydrology, Vol. 270 (1–2), pp. 105–134.

Flörke, M., Bärlund, I., Kynast, E. (2012): Will climate change affect the electricity production sector? A European study. Journal of Water and Climate Change 3 (1): 44-54.

Flörke M., Kynast E., Bärlund I., Eisner S., Wimmer F. and Alcamo J. (2013): Domestic and industrial water uses of the past 60 years as a mirror of socio-economic development: A global simulation study. Global Environ. Change 23: 144-156.

Hantulga (2009): Mongolia Gets $300M for Agriculture. Business Mongolia, March 18, 2009. http://www.businessmongolia.com/agriculture/mongolia-gets-300m-for-agriculture/#more-2993.

Hofmann, J., Rode, M., Theuring, P. (2013): Recent developments in river water quality in a typical Mongolian river basin, the Kharaa case study. Proceedings of the IAHS-IAPSO-IASPEI Assembly, Gothenburg, Sweden, July 2013, IAHS Publ. 361, 123-131.

Karthe D., Chalov S., Malsy M., Menzel L., Theuring P., Hartwig M., Schweitzer C., Hofmann J., Priess J., Shinkareva G. and Kasimov N. (2014): Integrating Multi-Scale Data for the Assessment of Water Availability and Quality in the Kharaa - Orkhon - Selenga River System. Geography, Environment, Sustainability 7(3): 65-86.

Korea Environment Institute (KEI) (2008): Integrated Water Management Model on the Selenge River Basin Status Survey and Investigation (Phase I), Tae Joo Park, Seoul, Korea, 443 pp.

Malsy M., aus der Beek T., Eisner S. and Flörke M. (2012): Climate Change impacts on Central Asian water resources. Adv. Geosci., 32, 77-83.

Malsy M., Heinen M., aus der Beek, T. and Flörke M. (2013): Water resources and socio-economic development in a water scarce region on the example of Mongolia. Geo-Öko 34(1-2): 27-49.

Menzel L., aus der Beek T., Törnros T., Wimmer F., and Gomboo D. (2008): Hydrological impact of climate and land-use change – results from the MoMo project. International Conference "Uncertainties in water resource management: causes, technologies and consequences". In: Basandorj, B. and Oyunbaatar D. (Ed.): IHP Technical Documents in Hydrology No. 1, UNESCO Office, Jakarta, 15–20.

O'Neill B. C. Kriegler E., Riahi K., Ebi K. L., Hallegatte S., Carter T. R., Mathur R. and van Vuuren D. P. (2014): A new scenario framework for climate change research: the concept of shared socioeconomic pathways. Climatic Change 122: 387-400.

Priess J. A., Schweitzer C., Wimmer F., Batkhishig O. and Mimler M. (2011): The consequences of land-use change and water demands in Central Mongolia. Land Use Policy, 28: 4-10.

Scharaw B. and Westerhoff T. (2011): A leak detection in drinking water distribution network of Darkhan in framework of the project IWRM in Central Asia, Model Region Mongolia. In: Гуринович, А.Д. (Ed.) (2011) Proceedings of the IWA 1st Central Asian Regional Young and Senior Water Professionals Conference, Almaty/Kazakhstan, 275-282.

Thorslund, J., Jarsjö, J., Chalov, S., Belozerova, E. (2012): Gold mining impact on riverine heavy metal transport in a sparsely monitored region: the upper Lake Baikal Basin 83 ENVIRONMENT case. Journal of Environmental Monitoring, 14 (10): 2780–92.

Törnqvist R., Jarsjö J., Pietroń J., Bring A., Rogberg P., Asokan S.M., and Destouni G. (2014): Evolution of the hydro-climate system in the Lake Baikal basin. Journal of Hydrology 519: 1953-1962.

Verzano K. (2009): Climate change impacts on flood related hydrological processes: Further development and application of a global scale hydrological model. Reports on Earth System Science 71–2009. Hamburg: Max Planck Institute for Meteorology.

van Vuuren D. P., Edmonds J., Kainuma M., Riahi K., Thomson A., Hibbard K., Hurtt G. C., Kram T., Krey V., Lamarque J.-F., Masui T., Meinshausen M., Nakicenovic N., Smith S. J. ,and Rose S. K. (2011): The representative concentration pathways: an overview. Climatic Change 109: 5-31.

Review of Long-term Satellite Data Series on Mongolia for the Study of Land Cover and Land Use

Martin Kappas[†], Tsolmon Renchin[‡], Selenge Munkhbayar[‡], Oyudari Vova[†‡], Jan Degener[†]

[†]Institute of Geography, Georg-August University Goettingen, Department of Cartography, GIS and Remote Sensing, Goettingen, Germany (Email: mkappas@gwdg.de)
[‡]NUM-ITC-UNESCO Remote Sensing and Space Science laboratory, Ulaanbaatar, Mongolia (Email: tzr112@psu.edu)

Abstract

The paper provides a short review about the availability of long-term remote sensing data time series over Mongolia. Further it focuses on remote sensing products that could be used for different ecological applications. Main focus is the availability of data time series for detection and assessment of Land Use / Land Cover change. The need of regional land data products is explained and their importance as input for Land Change models (LCM's) is highlighted. Additionally various operational applications over Mongolia based on satellite remote sensing data are presented.

Keywords – Long-term satellite data, operational remote sensing over Mongolia, remote sensing applications

Introduction

Remote sensing data are some of the most effective input data for Land Change Models (LCM's) [1]. In particular, multispectral and hyperspectral space-borne and airborne data are widely used to study changes in land use and land cover. Further different natural and anthropogenic processes including fire detection, snow mapping, and grassland / rangeland vulnerability are mapped and evaluated by remote sensing data.

Generally spoken there is a need for "Regional Land Data Products for Energy Budget and Water Cycle Trends and Processes in the future ..." (Source: ISLSCP: Int. Satellite Land-Surface Climatology Project; 2009). The major task is to produce consistent research quality data sets complete with error descriptions of the Earth's energy budget and water cycle and their variability and trends on interannual to decadal time scales, and for use in climate system analysis and model development and validation. Therefore long-term satellite data are important input data for Land Change Models (LCM's) that link patterns and processes across multiple scales. The output of these LCM's is severely dependent on the quality of the input data (mostly remotely sensed data). Whereupon Land Cover / Land Use Change (LCLUC) is an interdisciplinary scientific theme within the ultimate vision is to develop the capability for periodic global inventories of land use and land cover from space, to develop the scientific understanding and models necessary to simulate the processes taking place, and to evaluate the consequences of observed and predicted changes. Therefore the next chapter takes a look on available satellite missions and their significance for broad environmental oriented applications over Mongolia.

Land Use / Land Cover relevant missions

The main missions to analyze land use and land cover from space can be separated into systematic and exploratory missions (see figure 1). Figure 1 shows a few examples of both groups. The systematic observations deal with observations of key earth system interactions whereas the exploratory missions focus on specific earth system processes (e.g. land degradation) and specific parameters (e.g. NDVI / LAI dynamics, drought indices, dzud events). Also new technology development belongs to the exploratory missions. Well known missions are the Landsat mission, the AVHRR (Advaced Very High resolution radiometer) or the MODIS (Moderate Resolution Imaging Spectroradiometer) mission which is a key instrument aboard the Terra (originally known as EOS AM-1) and Aqua (originally known as EOS PM-1). Terra's orbit around the Earth is timed so that it passes from north to south across the equator in the morning, while Aqua passes south to north over the equator in the afternoon. Terra MODIS and Aqua MODIS are viewing the entire Earth's surface every 1 to 2 days, acquiring data in 36 spectral bands. These data will improve our understanding of global dynamics and processes occurring on the land.

MODIS, AVHRR and SPOT Vegetation data are playing a vital role in the development of validated, global, interactive Earth system models able to predict global change accurately enough to assist policy makers in making sound decisions concerning the protection of our environment.

Figure 1. Examples of Systematic and Exploratory missions relevant over Mongolia (Source: Garic Gutman, oral presentation 2014 Ulaanbaatar).

The mentioned satellite data from AVHRR, MODIS and SPOT Vegetation provide coarse resolution information about the earth surface. The pixel resolution in correspondence to the data product varies between 250m (MODIS) to 1, 4 or 8 km (SPOT, AVHRR). AVHRR offers the longest available satellite based data set on earth. The important AVHRR based NDVI3g and LAI3g data sets are available for entire Mongolia with high product continuity. These products are delivered with elaborated documentation and product validation [2]. Figure 2 gives an overview of current available satellite data sets over Mongolia.

Sensor	Satellite	Overpass/ Orbit Frequency	Data Source (terrestrial data)	Data Record (years)	Spatial Resolution(s)	Processed Time Step	Latency
AVHRR	NOAA series	Daily	USGS/EROS	1989-present	1 km	1-week, 2-week	~24 hours
AVHRR	NOAA series	Daily	Global Land Cover Facility	1982-2006	8 km	Twice monthly	N/A
MSS	Landsat 1-5	18 days	USGS/EROS	1972-1992	79 m	Distributed by scene	N/A
TM	Landsat 4-5	16 days	USGS/EROS	1982-2011	30 m	Distributed by scene	N/A
ETM+	Landsat 7	16 days	USGS/EROS	1999-present	30 m	Distributed by scene	~1-3 days
Vegetation	SPOT	1-2 days	VITO	1999-present	1.15 km	10-day	~3 months
MODIS	Terra	1-2 days	LPDAAC	2000-present	250 m, 500 m, 1 km	8-day, 16-day	~7-30 days
MODIS	Aqua	1-2 days	LPDAAC	2002-present	250 m, 500 m, 1 km	8-day, 16-day	~7-30 days
eMODIS	Terra/ Aqua	1-2 days	USGS/EROS	2000-present	250 m, 500 m, 1 km	7-day	~15 hours, 7 days[6]

Figure 2: Most important satellite missions over Mongolia. In future many more sensors for ecosystem analysis like Sentinel, EnMAP and others will be available. Sentinel 2a was just launched in June 2015 and Sentinel 2b will follow in 2016.

Land Monitoring at Moderate Resolution

Moderate spatial resolution sensors (100–300 m) such as MODIS/MERIS with frequent (daily to weekly) or coarse resolution sensors (1000 m) such as NOAA AVHRR, SPOT Vegetation or SEASAT with their sensitive and unbiased observations of vegetation properties such as the Fraction of Absorbed Photosynthetically Active Radiation (FAPAR) or Leaf Area Index (LAI) deliver fundamental indicators for environmental assessments and have already been recognized as 'Essential Climate Variables' (ECV's) by the Global Climate Observing System (GCOS). These measurements derive quantitative information about the environment and are useful for assessments. Indicators such as FAPAR or LAI replace dimensionless indices such as the Normalized Difference Vegetation Index (NDVI) [3, 4]. These data sets are completely available over Mongolia, but they need intensive validation by crosschecking with ground truth data.

Landsat satellite family presents the most important information resource in the lower moderate resolution of 30m pixel size for regional studies. Landsat data are accessible free of charge at USGS and the Landsat Data Continuity Mission (LDCM; Landsat-8 was launched Feb 11, 2013) is an important information source for further environmental studies over Mongolia. But Landsat observations are also insufficient and international cooperation is needed to fill the gaps to provide continuous data series. A challenging approach is the

WELD project (Web-enabled Landsat data) that is using all clear pixels by com-
positing to derive a cloudless mosaic over the landscape.

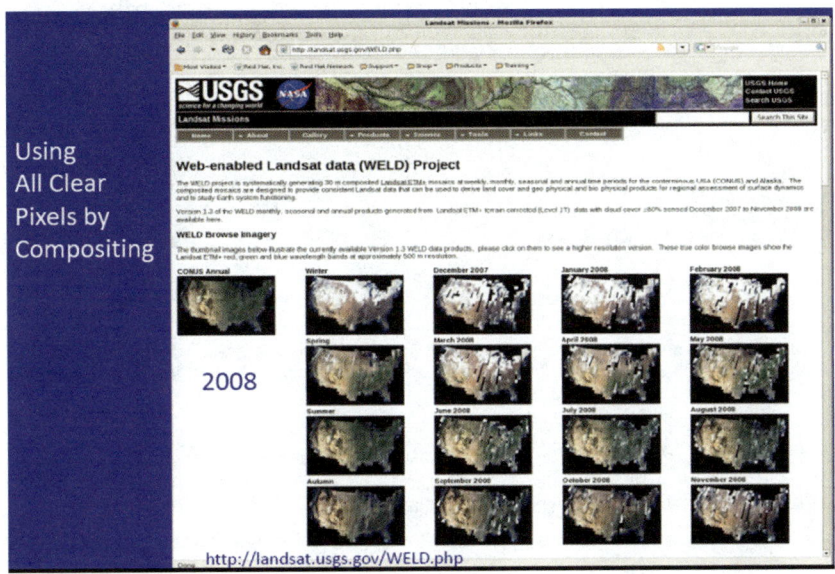

Figure 3: Web-enabled Landsat data project (WELD), see: http://landsat.usgs.gov/WELD.php

The Landsat data are available from 1972 (L1) up to date. Our review could
discover 14433 Landsat images in the archive (status: August 2014) with a
cloud cover (<10%) over Mongolia. A general problem is the image size of
(185x185 km) that no consistent mosaic for Mongolia is producible because of
a 16 days repetition cycle to monitor the same area. Cloud cover, atmospheric
influences, sun angle changes and many other influences require comprehensive calibrations. Therefore Landsat data are better for regional studies than
for entire Mongolia. According to the specific Landsat mission we have different
tiling over Mongolia (varies between 23-26 tiles; e.g. WGS-2 Path:130/Row28;
Landsat 8: 13; Landsat 7:   15; Landsat 4-5 TM: 77; Landsat 4-5 MSS: 0;
Landsat 1-3 MSS (WMS-1): 0; total tilling: 105). The Landsat data are freely
available under http://earthexplorer.usgs.gov/.
Figure 4 shows an example of Land cover map derived from satellite data and
ground truth data along a transect over Mongolia (Xilin Gol Transect).

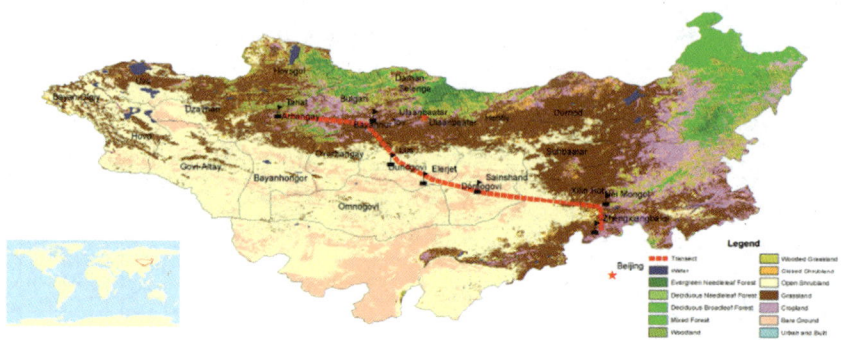

Fig.1 Land Cover of Mongolian Plateau based on AVHRR Data and Tariat – Xilin Gol Transect

Figure 4. Example of remote sensing based Land Cover Map over Mongolia. Source: Yunfeng Hu, Yifang Ban, Qian Zhang; Department of Urban Planning & Environment Royal Institute of Technology Stockholm, Sweden 2008

Examples of operational use of satellite data over Mongolia

A Satellite Observation System is available since 1970, where Mongolia has received information and images from the Polar orbit satellites. A digital information station was installed during 1986 -1988. An Arc/INFO GIS package on Sun Sparc Workstation was installed in 1994. Cooperation agreement with NASA was signed in 1993 to use satellite SEASTAR. With the satellite-aided observation, the monitoring of forest fires and bushfires became possible. Since 2007, Mongolia has been receiving satellite images from MODIS which increased monitoring quality significantly. Based on MODIS data Mongolia developed several operational monitoring tools that deal as an information source for decision making. One important operational tool is the Mongolia Livestock Early Warning System (Mongolia LEWS). During the period from 1999 to 2002, Mongolia experienced a series of droughts and severe winters that diminished livestock numbers by approximately 30% countrywide. In the Gobi region, livestock mortality reached as much as 50%. Due to these extreme events and its impact on pastoral livelihoods, the USAID mission in Mongolia and the Global Livestock-CRSP (GL-CRSP) initiated the Gobi Forage program with the goal of transferring Livestock Early Warning System (LEWS) technology to Mongolia. The Livestock Early Warning System technology combines near real-time weather, computer modeling, and satellite imagery to monitor and forecast live-stock forage conditions so that pastoralists and other decision makers get information for timely decision making. Three major activities have been conducted including: 1) infusion of forage monitoring technology to assess regional forage quantity; 2) development of nutritional profiling technology to assess forage quality, and 3) information delivery and outreach (source: http://glews.tamu.edu/mongolia/pagesmith/2). Figure 5 shows an example of the operational monitoring system for a 60 day forage forecast.

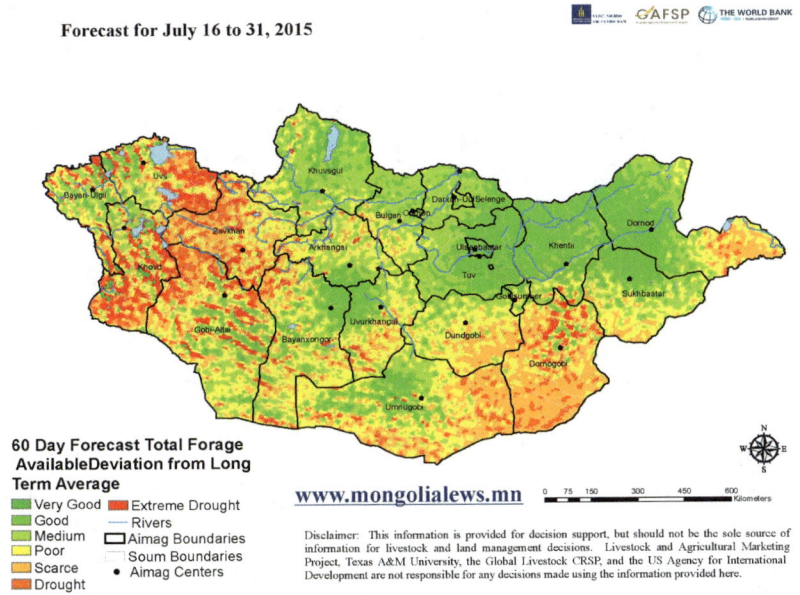

Figure 5. Example of a 60 day forecast total forage over Mongolia based on Mongolia-LEWS operational system. (Source: http://www.mongolialews.net/images/filecabinet/forage-maps//2015-05-31/2015-05-31-en-go-mo-bg-forage-deviation-60day.jpg)

On the base of the Mongolia-LEWS many other applications are possible. The Mongolia-LEWS presents a good example of operational use of long-term satellite data available over Mongolia. Based on MODIS data not only forecasts of amount and quality of pastures in Mongolia are possible but also other important questions can be solved with the help of this operational system. Another important issue is the derivation of a Dzud-Index for better adaptation to severe winter conditions. A snow index map for entire Mongolia is also available.

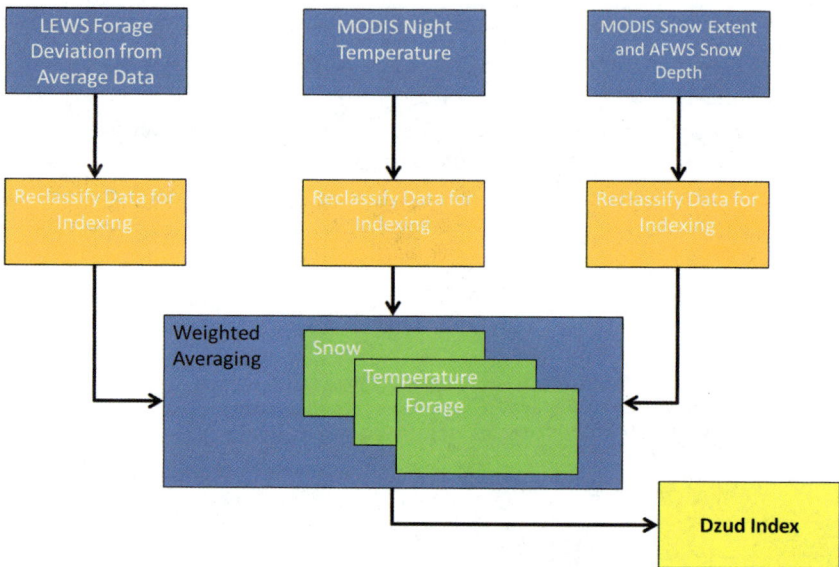

Figure 5. Derivation of a Dzud-Index based on MODIS stellite data and additional meteorological data from Mongolia-LEWS.

Conclusion

Long-term satellite data over Mongolia offer many possibilities to develop ecosystem assessments and can be used to create operational systems for many different applications. Looking back on 40 years research about biomass with Remote Sensing and ground truth data, estimates of pasture biomass amount for livestock fodder and pasture carrying capacity over whole territory of Mongolia is possible. In average the following values for spring potential biomass can be derived from this research: 27-50 gm^2 in the forest steppe, 15-33 gm^2 in the steppe, 5-13 gm^2 in the Altai Mountains and 3-6 gm^2 in the Gobi desert. All Remote Sensing based studies showed that in past 40 years the total pasture carrying capacity was drop down by 27 % because of biomass decrease. Based on current and future satellite data there is a strong need of applications in the near future such as applications for disaster monitoring on pasture, drought and other disasters, generating products for heavy snow (dzud), forest and steppe fires and dust storms.

The biggest challenge in the framework of using long-term satellite data over Mongolia is the integration of these data with other environmental data into a unique GIS-based system to build up an integrated Land Data Assimilation System for Mongolia (LDAS-Mongolia) to push political decisions and provide decision support.

References

D.G. Brown, R.Walker, S. Manson, K. Seto. "Modeling Land Use and Land Cover Change". In: Land Change Science. Remote sensing and Digital Image Processing Volume 6, pp. 395-409, 2004.

M. Kappas, P. Propastin, J. Degener, T. Renchin, "Inter-Comparison and Evaluation of the Global LAI Product (LAI3g) and the Regional LAI Product (GGRS-LAI) over the Area of Kazakhstan. Remote Sensing 7(4), pp. 3760-3782, 2015.

J. Chen, S. Wan, G. Henebry, J. Qi, G. Gutman, G. Sun, M. Kappas (eds.): "Dryland East Asia: Land Dynamics Amid Social and Climate Change". 415 p., De Gruyter. 2013

M. Kappas, P. Propastin "Monitoring and Assessment of Drylands Ecosystems with Remote Sensing", In: J. Chen, S. Wan, G. Henebry, J. Qi, G. Gutman, G. Sun, M. Kappas (eds.): "Dryland East Asia: Land Dynamics Amid Social and Climate Change", De Gruyter/Higher Education Press, pp. 309-349, 2013.

Drivers of Land degradation in Umnugobi Province

Tsolmon Renchin[‡], Martin Kappas[†], Selenge Munkhbayar[‡], Oyudari Vova[†‡], Jan Degener[†]

[‡]NUM-ITC-UNESCO Remote Sensing and Space Science laboratory, Ulaanbaatar, Mongolia (Email: tzr112@psu.edu)
[†]Institute of Geography, Georg-August University Goettingen, Department of Cartography, GIS and Remote Sensing, Goettingen, Germany (Email: mkappas@gwdg.de)

Abstract

Remote Sensing and GIS were used to monitor interactions and relationships between land use and land cover changes in the regional ecology area of Umnugobi province (South Gobi).This study aims at determining the land degradation conditions in 15 soums (administrative units) of the study area, Umnugobi province. Using GIS processing of data climate drivers (precipitation, air temperature) vegetation data and socio-economic drivers (livestock numbers, population figures, mining activities) were analyzed. We focused on developing a modeling approach for monitoring land degradation using GIS and Remote Sensing tools by integrating natural and socio-economic data. The Moderated Soil Adjusted Vegetation Index (MSAVI) from SPOT/VEGETATION was used to determine vegetation cover change for the period 2000 to 2013. Landsat data for the years 2000, 2010 and 2013 were analyzed to derive and classify "hot spot" areas of land degradation. GIS conditional functions were used for mapping and analyzing climate and socio-economic driving factors, both of which affect land degradation. Conditional functions such as MAP-Algebra from ArcGIS were developed using ground truth data and data from National Statistics. Our study documents that 60 percent of the study area is affected by land degradation caused by human and climate drivers.

Keywords – socio-economic change, mineral resources, grassland degradation, land degradation monitoring

Introduction

The Gobi Region is an enormous area in Mongolia, sparsely populated and rich in mineral resources. Mongolia is susceptible to climate change due to its geographic location, vulnerable ecosystems and an economy that is highly dependent on seasonal climates (e.g. pastoral systems). In the past 40 years, climate change and other anthropogenic activities have had a significant impact on the Mongolian ecosystem, resulting in desertification, increased occurrences of drought, water source depletion, and a decrease in biological diversity as well as affecting the well-being of local communities. The annual average temperature in Mongolia increased by 2.14°C between 1940 and 2008. When comparing this rise to the global annual average temperature increase, a rise of 0.85°C during the period 1880-2012, climate change is occurring rapidly in Mongolia and having a strong impact on melting glaciers.

Pastoral systems, where humans depend on livestock, exist largely in arid and semi-arid ecosystems in Mongolia, where the climate is highly variable. In many ways, pastoral livestock systems are closely adapted to climatic variability. However, there has been little research done on how past climate records across extensive areas, such as the Gobi desert of southern Mongolia and northern China, could serve as a baseline to identify potential climate and environmental conditions (Hulle et al. 2010; Felauer et al. 2013). Calculations based on the methods outlined in the UN's Convention to Combat Desertification (UNCCD) showed that approximately 90 percent of pasture land in Mongolia lies within a vulnerable region that is susceptible to desertification and land degradation.

Several studies have dealt with climate controls on primary vegetation production in the dry land areas of Mongolia. These studies focused on the water component as the major climate constraint as identified by Nemani et al. (2003). Ni (2003) and Li et al. (2007) showed that relationships existed between vegetation production and both precipitation and evapotranspiration in Mongolia. Miyazaki et al. (2004) and Munkhtetseg et al. (2007) demonstrated the influence of climate on vegetation production in the growing-season. They both showed that precipitation in July had the greatest influence on vegetation production. Furthermore, other studies (Zhang et al. 2005; Munkhtetseg et al.

2007) concluded that air temperature had a negative influence on vegetation production, especially in dry regions.

Geist and Lambin (2004) found six common human and biophysical factors associated with the proximate causes of desertification. These factors included demographic factors, economic factors, technological factors, policy and institutional factors, climate factors and cultural factors. Each of these factors was associated with four common proximate causes of desertification namely: agricultural activities, infrastructure extension, wood extraction, and increased aridity.

(Kappas and Propastin 2008 a, b) noted that drivers can determine both anthropogenic (e.g. demographic change) and natural forces. Human activities can be changed by socio-economic factors, which increase or diminish pressures on the environment. Human based drivers can be altered by policy and societal influences or societal responses. Natural based drivers (e.g., climate variability) cannot be controlled by people directly, but must be considered for future land management and are more or less policy related.

In general only a limited amount of research exists on drivers of land degradation in Mongolia. In order to determine land degradation in a specific area such as in the Mongolian Gobi region, there is a demand to develop a modeling approach applying GIS and Remote Sensing.

The objective of this research is to assess current trends in driving factors affecting land degradation in Umnugobi province. There is an urgent need to assess the ways in which people interact with the environment in this study area. The study contributes to modeling land use change at a regional level. Remote Sensing and GIS assessed relationships between land use and land cover changes in the regional area by monitoring how local people were affected and how they were interacting with the landscape. The climate and natural driving factors (precipitation, air temperature, and vegetation condition) and economically driven factors (population density, mining activities) were analyzed. Each of these factors can cause land degradation. Our research approach shows that any other driving factors, such as policy or cultural aspects, can also be estimated.

Study Area

Mongolia is a unitary state and divided administratively into 21 aimags (provinces) and the capital city UlaanBaatar. The aimags are subdivided into 230 soums. The study area is Umnugobi

Mongolian desert and desert steppe: Here the mean annual precipitation is about 100 to 125 mm. The desert zone of Mongolia is the northern edge of the Central Asian Desert. Sand dunes, drifting sand, and salt marshes are common. Vegetation is scarce in this zone. Shrubs grow intensively during the rainy seasons. The Gobi remains a region where nomadic pastoralism has been practiced for millennia and continues to provide livelihoods for 30% of the rural population (Sternberg et al. 2009; Ulambayar and Fernandez-Gimenez 2013). Herders, communities and local governments have traditionally depended on shallow groundwater for basic needs (5-10 liters/person per day; UNEP 2011) and livestock (43 million animals in 2013; National Statistical Yearbook 2013), because of the feasibility and lower cost of access compared to deep water sources.

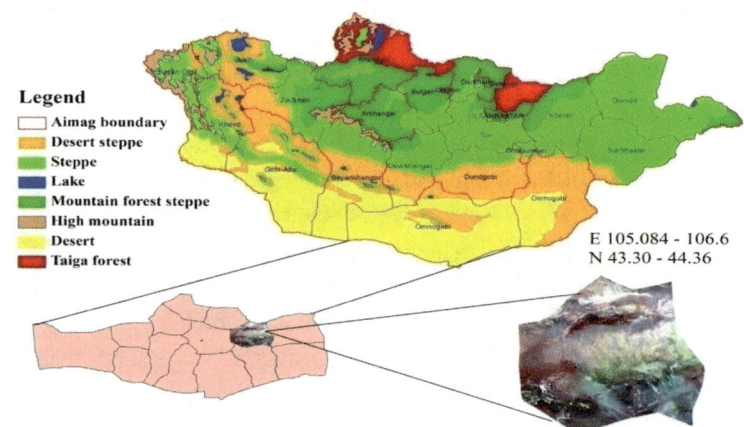

Figure 1. Study area – Umnugobi province

OyuTolgoi, Rio Tinto's enormous copper-gold mine and TavanTolgoi, the world's largest coking coal deposit (Reuters 2011), sit in the middle of Umnugobi Province and are part of the study area. Local populations that lack running water, depend on the hauling of water from mechanized wells (in towns) and shallow hand wells (in the countryside) inside the study area. Summing up, we can note that the study area offers a variety of land degradation drivers for a comprehensive study.

We used remote sensing data from SPOT 4 VEGETATION with 1km resolution for the time period from April to October for the years 2000 to 2013.GIS data (climate data, socio-economic data) and LANDSAT ETM+ for the years September 2000, 2010 and 2013 were used in this research at the "hot spot"-level. The time series of SPOT 4VEGETATION RED and NIR channel data was used for calculation MSAVI and monitoring vegetation change inside the entire province. Additionally, statistical data for socio-economic development in the 15 soums of Umnugobi province and climate data (based on local climate stations in the province) were integrated into GIS for an explicit analysis.

For the 15 soums (Bayandalai, Bayan-Ovoo, Bulgan, Gurvantes, Mandal-Ovoo, Manlai, Nomgon, Noyon, Sevrei, Khanbogd, Khanhongor, Hurmen, Tsogt-Ovoo, Tsogttsetsii, Dalanzadgad) data on population, number of households/soum, number of herder households/soum, total number of herders, composition of livestock such as number of horses, cattle, sheep, goats or camels were collected and mapped inside the GIS.

Methodology

With regard to monitoring land degradation with remote sensing, we applied MSAVI indexes to monitor vegetation change over time in the years 2000 to 2013. In order to analyze socio-economic and climate factors, conditional function MAP-Algebra from ArcGIS was applied. Many analytical tasks in the Grid rely on Boolean and conditional statements. At the simplest level we used the CON function inside ArcGIS to compute the influence of driver combinations.

Output maps from remote sensing and conditional function maps from GIS were compared with each other. For validation, we used classifications derived from Landsat data to derive hot spot areas (e.g. areas below a certain MSAVI-threshold). In a first step we calculated MSAVI for Spot 4 Vegetation data at

1km resolution. Next we assessed land cover changes between the years 2000 and 2013 using MSAVI. Then areas with the greatest amount of land cover change were identified and assessed with higher resolution Landsat imagery for the years 2000 and 2013.

The Landsat imagery was used to zoom into those areas that were found to have the greatest amount of potentially human-induced change. Considering the spectral domain, spectral bands (Near Infrared 0.78-0.89 µm, Short wave Infrared1.58-1.75 µm) and algorithms1-2 were selected for vegetation mapping.

Huete (1998) suggested a new vegetation index, which was designed to minimize the effect of the soil background, which he called the soil-adjusted vegetation index (SAVI) (1) developed from an iterated version of this vegetation, which is called MSAVI (2)

$$SAVI = \frac{NIR - RED}{NIR + RED + L} * (1 + L)$$

(1)

$$MSAVI2 = \left[2NIR + 1 - \sqrt{(2NIR + 1)^2 - 8(NIR - RED)} \right] / 2$$

(2)

Derived MSAVI values vary from 0.03 to 0.09 for the period 2000 to 2013 (figure 2) which means that there is only a sparse vegetation cover on the ground. The spatial distribution of vegetation is indicated in the MSAVI maps by red colors (high vegetation condition), while dark blue and dark pink colors indicate low vegetation conditions (figure 3).

The main climatic drivers are precipitation development and temperature increase over the growing period. Figure 4 shows that precipitation had an influence on vegetation production; while air temperature had a negative influence on vegetation in the study area (figure 5).

The figure 6 describes land cover classification for the Tsogttsetsii soum of Umnugobi province based on Landsat data. This soum was chosen as a hot spot area.

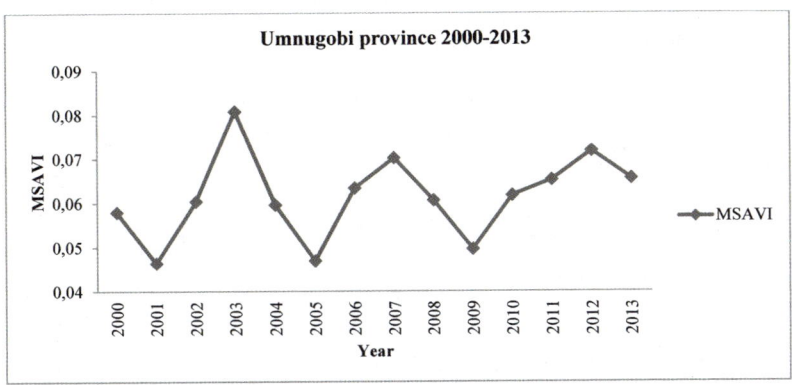

Figure 2: MSAVI values for the period 2000 to 2013Data

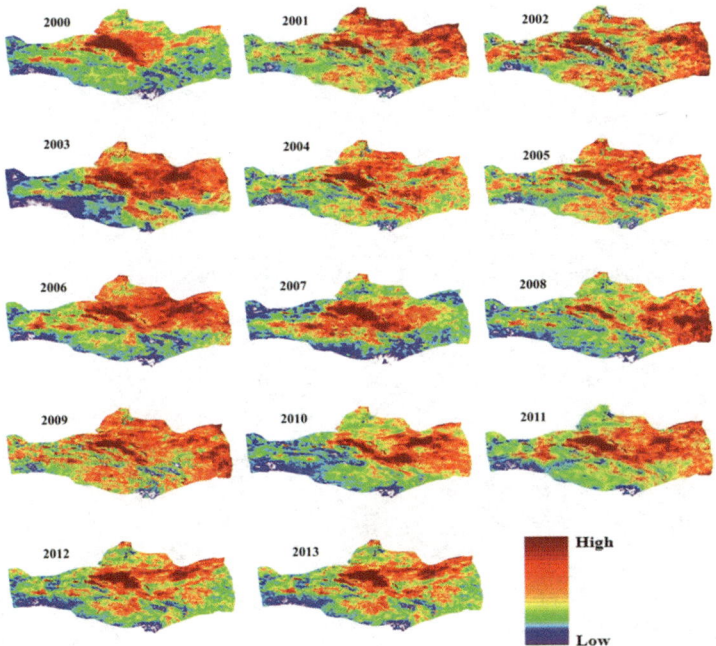

Figure 3. MSAVI change map for the period 2000 to 2013

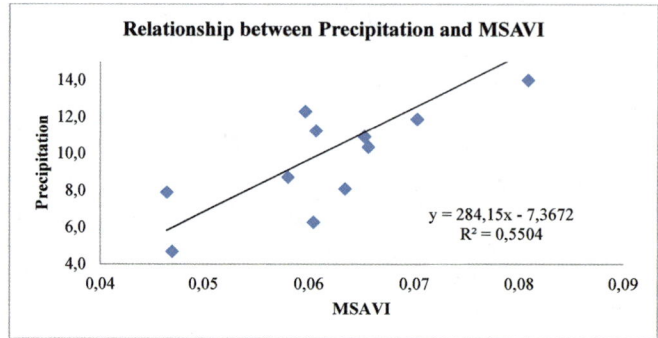

Figure 4. Relationship between Precipitation and MSAVI in the Umnugobi province

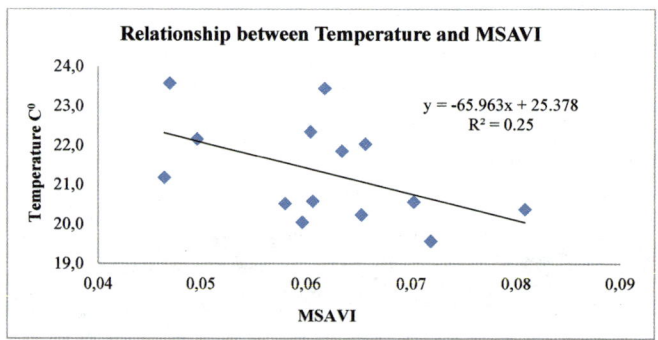

Figure 5. Relationship between Temperature and MSAVI of the Umnugobi province

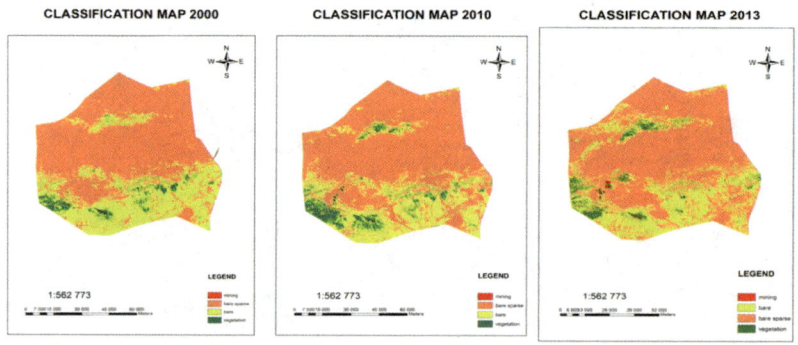

Figure 6. Land cover classification for in Tsogttsetsii soum,Umnugobi province based on Landsat data for years, 2000, 2010 and 2013

GIS analysis

Using ground truth measurement, statistical data, and expertise working in the study area, we developed degradation maps using Map Algebra function in GIS. The ESRI ArcGIS software provides access to Map Algebra functions and operators. The conditional function (CON) was employed in the analysis of the impact of a number of socio-economic factors. Map Algebra functions operate on data that is in raster/grid format.

The basic form of a Con function statement can be seen below.

I.   *Con(<condition>, <true_expression>, {false_expression})*

In formula I above, <condition> is a conditional expression that is evaluated for each cell in the participating raster datasets. If the condition is true, <true expression> identifies the value to be used to compute the output cell value. If none of the results of the evaluations of the conditional statements is true, a value or expression can be applied to the cells through the {false_expression} optional argument. (ESRI, 2008)

The impact of socio-economic factors was defined by looking at the number of animals (e.g. goats), the population figures, the number of mining activities measured as increasing mining space, and climate impact (precipitation, and temperature change). All of these factors were used as conditional statements (formulas II-VII). Con-function statements can be nested. Formulas II-VII each show one nested Con-function statements. The result of processing each formula was a new output raster dataset. From the field experience we selected the most degraded areas. Table 1 shows socio-economic average data for all soums over the years 2000-2013. The four soums Bayan-Ovoo, Kanbogd, Gurvantses, Tsogtsestsii were selected as being the most degraded area in the province (Table 1). Average data from these soums were used to construct degradation function as follows

II. con ([Livestock number] > 165677, 1, 0)

III. con ([MSAVI] < 0.05, 1, 0)

IV. con ([Temperature] > 22, 1,0)

V. con ([Precipitation] < 8, 1, 0)

VI. con ([Population] > 4358, 1, 0)

VII. con ([Mining area >2 km², 1, 0)

In formula II, if the number of livestock was greater than 165677, then the output was assigned a value of 1. A value of 1 is defined as a signifier of land degradation. If these conditions were not met, the output was assigned a value of 0, and then land degradation was not significant. In formula 3 if MSAVI was less than 0.05 then the output was assigned a value of 1. A value of 1 signified land degradation. In formula 4, if the temperature was higher than 22°C, during the vegetation season (June to September), the output was assigned a value of 1 which signified land degradation. In formula 5 if the precipitation was less than 8 mm per vegetation period, then the output was assigned a value of 1. A value of 1 signified land degradation. If these conditions were not met, the output was assigned a value of 0, and then land degradation was not significant.

In formula 6, if the population was greater than 4,358, then the output was assigned a value of 1, which signified land degradation. If these conditions were not met, again the output was assigned a value of 0. In formula 7, if the mining area is greater than 2 km², then the output was assigned a value of 1, which signified land degradation. If these conditions were not met, again the output was assigned a value of 0 and then land degradation was not significant. Finally, the six output raster datasets were summarized into one raster output dataset by summing the values at each cell location. Possible output cell values ranged from 0 – 6, illustrating different intensities of land degradation conditions. If an output cell indicates a value of 6 that means the 6 drivers have an influence on land degradation. If an output cell indicates 0 it means there is no land degradation detectable according to this simple valuation method.

Summary of the con-functions

The conditions on each GIS maps were graded on their relative land degradation (0=no land degradation in yellow colors through 6=land degradation in red color). Figure 7 shows output conditional land degradation maps.

Socio-economic data for the years 2000 – 2013 were processed. In each case the output was compared to ground truth data and MSAVI data derived from SPOT Vegetation for the years 2000-2013 during the vegetation growing season (April to October).

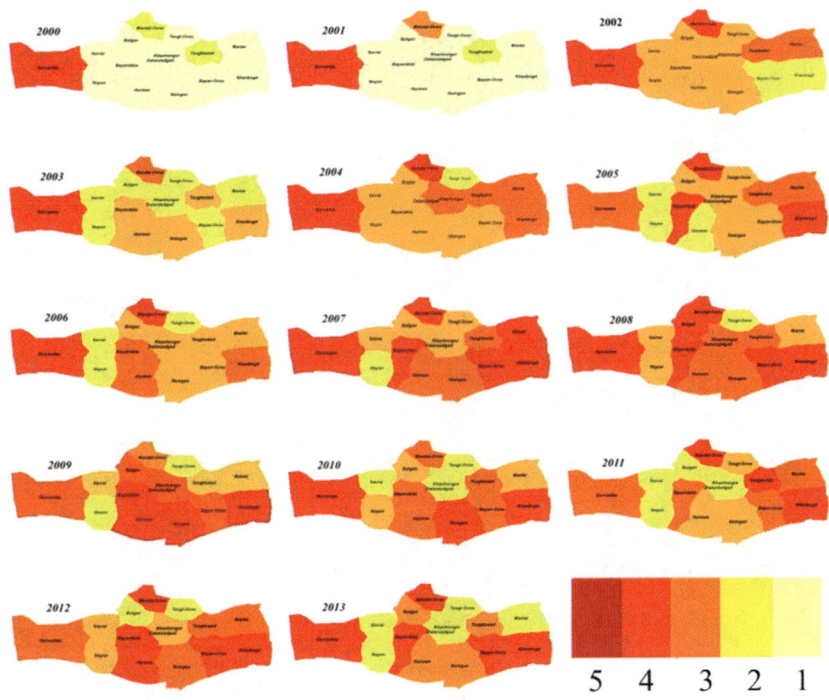

Figure 7. Land degradation condition map for the period 2000 to 2013

Table 1. Socio-economic data for the analysis 2000-2013

Soum names	Population	Livestock	MSAVI	Temperature C⁰	Precipitation mm	Mining area km2 2000 year	Mining area km2 2013 year
Bayandalai	2,148	176,320	0.06	20.5	11.1		15.2
Bayan-Ovoo	2,148	133,233	0.04	22.4	8.9		692.02
Bulgan	2,187	145,891	0.07	21.5	6.1		
Gurvantes	4,465	168,362	0.05	21.7	6.8	1.55	839.58
Mandal-Ovoo	1,687	176,309	0.03	22.3	8.7	0.82	34.65
Manlai	2,522	196,692	0.04	19.8	6.8		14.98
Nomgon	2,624	240,139	0.05	20.7	7.9		115.10
Noyon	1,388	113,257	0.07	23.0	5.8		390.61
Sevrei	2,009	154,388	0.06	23.0	7.6		
Khanbogd	4,712	259,781	0.06	22.4	10.4		1796.68
Khanhongor	2,113	172,134	0.06	22.0	9.9		
Hurmen	1,644	151,820	0.07	20.7	6.9		29.66
Tsogt-Ovoo	1,644	100,974	0.05	24.3	10.7		
Tsogttsetsii	6,108	101,332	0.05	23.3	12.8	2.62	1186.11
Dalanzadgad	21,581	78,187	0.06	20.0	9.9		

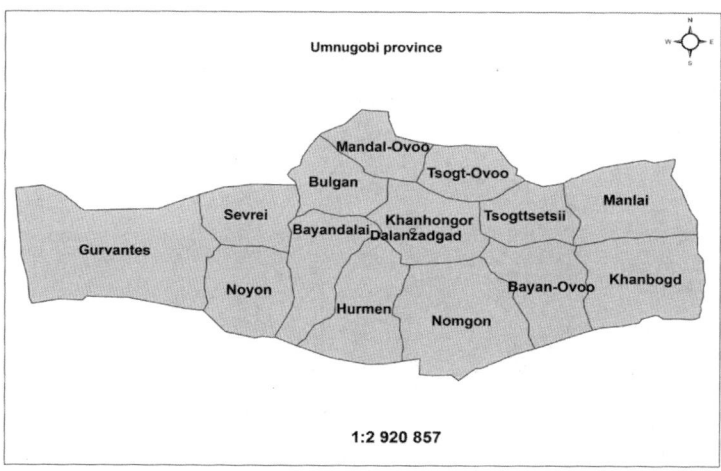

Figure 8. Soums of the Umnugobi province

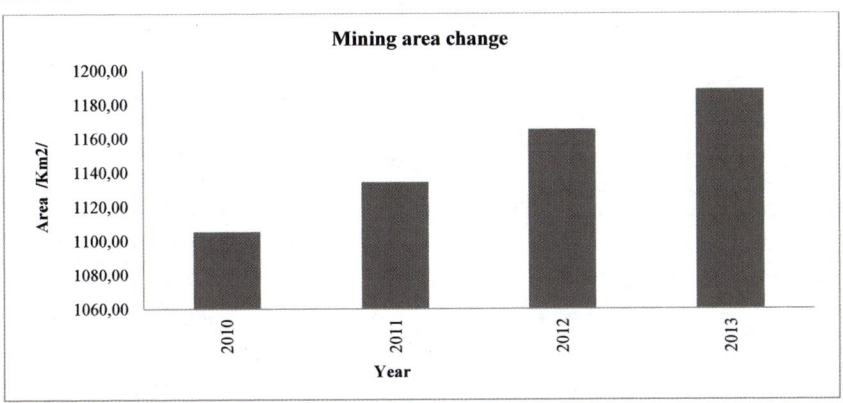

Figure 9. Mining area change in the Tsogttsetsi soum

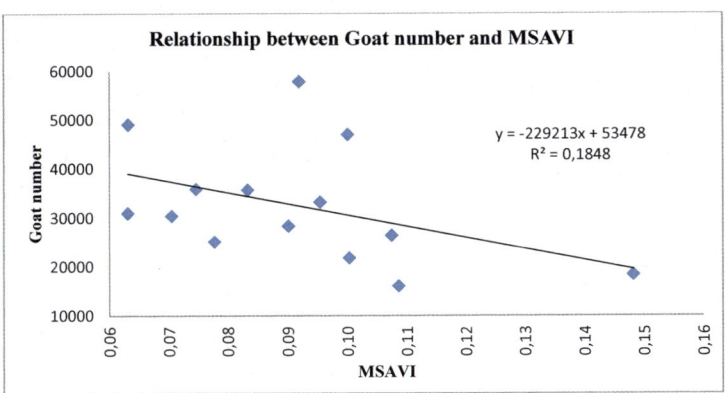

Figure 10. Relationship between Goat number and MSAVI in the Tsogttsetsii soum

Results and discussion

Vegetation index (MSAVI) from SPOT data was applied in this area in order to determine vegetation cover change in the time period from 2000 to 2013. As shown on figure 2 MSAVI values vary from 0.03 to 0.09. There is virtually little change over the 13 years for the study area. By means of MSAVI variation alone we cannot assess the land degradation process. The result indicates that the vegetation is not a main driving factor for land degradation in the study area. GIS analysis was completed for 15 soums of the Umnugobi province to determine which soum is affected most by driving factors (figure 8). Raster condition maps using the factors (number of goats, population figures, number of mining activities, climate impact such as precipitation and temperature change) for the time period from 2000 to 2013 have the outcome that "Gurvantes" is the soum with the highest land degradation in the study area. Another result shows that by the year 2000 only one soum (Gurvantes) had an indicative value of land degradation that comes from a larger number of goats during this time that caused overgrazing there in comparison to the other soums. In 2013 there were 5 soums most affected by all drivers and 4 soums were affected by some drivers namely expansion of mining areas, total population and livestock numbers. The soum "Tsogtsetsii" was not affected by any driving factors from 2000 to 2001. Since 2002 there is land degradation steadily increasing in this soum with an increasing population and number of animas number and the expansion of mining areas. Tsogtsetsii soum surveyed by a Landsat based land cover change analysis for the years 2000 and 2013 showed the greatest amount of land cover change today. The main driver is the expansion of the mining area in Togtstestii by around 9200 km² from 2000 to 2013 (figure 9). Increasing opportunities in the mining sector, despite increasing animal husbandry have a negative impact on the environment in this soum. The factors having the most impact in this soum were goat numbers and mining activity numbers. The simple map algebra approach allows us to determine which driving forces exist in the most degraded areas. These results should be useful for identifying "hotspots" of land degradation in Mongolia. Indeed this study enables us to infer that such land degradation "hotspots" are in fact the result of changes in climate or in socio-economic drivers such as livestock, population, mining activities or other impacts.

A general statement from our statistical analysis of the livestock numbers over the growing season is that the number of goats is significantly greater than that of other animal groups. An increasing number of goats provoke overgrazing on the one hand but on the other hand goats are economically important in the area because of cashmere production and the value creation chain (figure 10). A total of 60 percent of the study area is affected by climate and socio-economic drivers. According to the National Report (2014) on Mongolia 77.8% of the Mongolian landscape has been degraded at some level. The areas where severe desertification occurred in the Umnugobi province are: the Khanbogd, Tsottsetsii and Manlai soums.

Due to increasing mining activities, there is an increased population caused by people migrating from the other parts of the country to these areas. This study contributes to the research which involves policy makers and stakeholders defining and negotiating relevant scenarios with participatory approaches in the local area as well as to the studies which link people to the environment. Finally, the study argues the this basic modeling approach can be used in other dry land regions in order to determine the precise driving factors of land degradation.

Acknowledgement

The authors and I would like to thank SPOT /VEGETATION data center for providing satellite data. I am also grateful to the DAAD Germany for supporting my research in Germany and to the members of the Department of Cartography, GIS and Remote Sensing, Institute of Geography at Gottingen University for hosting me as a researcher.

References

Convention on biological diversity The 5th National Report of Mongolia, Ulaanbaatar Mongolia 2014

Geist H.J., Lambin, E.F.: Dynamic causal patterns of desertification. Bioscience, 54, 817-829, 2004.

Heute, A.R.: A soil-adjusted vegetation index (SAVI). Remote Sensing of Environment, 25, 295-309, 1998.

Li, S., Asanuma, J., Kotani, A., Davaa, G. and Oyunbaatar, D.: Evapotranspiration from Mongolian steppe under grazing and its environmental constraints. Journal of Hydrology133-143, 2007.

Lambin E.F., Turner II, B.L.,Geist, H.J., Agbola, S.B., Angelsen, A., Bruce, J.W.,Coomes, O.T., Dirzo, R., Fischer, G., Folke, C., George, P.S., Homewood, K.,Imbernon, J. Leemans,R., Li, X., Moran, E.F., Mortimore, M., Ramakrishnan, P.S.,Richards, J.F., Skanes, H., Steffen, W.,Stone,

G.D., Svedin, U., Veldkamp, T.A.,Vogel, C. and Xu, J.: The causes of land-use and land-cover change – Moving beyond the myths. Global Environmental Change: Human and Policy Dimensions,11, 261-269. 2001

Munkhtsetseg, E., Kimura, R., Wang J. and Shinoda, M.: Pasture yield response to precipitation and high temperature in Mongolia. Journal of Arid Environments, 70, 94110, 2007.

Miyazaki, S.,Yasunari, T.,Miyamoto,T., Kaihotsu, I., Davaa, G., Oyunbaatar,D., Natsagdorj,L. and Oki, T.: Agro meteorological conditions of grassland vegetation in central Mongolia and their impact for leaf area growth. Journal of Geophysical Research, 109, D22106. 2004.

Ni, J.: Plant functional types and climate along a precipitation gradient intemperate grasslands, north-east China and south-east Mongolia. Journal of Arid Environments, 53, 501–516, 2003.

Propastin, P. A., Kappas, M.: Inter-annual changes in vegetation activities and their relationship to temperature and precipitation in Central Asia from 1982 to 2003. Journal of Environmental Informatics. Vol. 12, Issue 2: 75-87, 2008a.

Propastin, P., Kappas, M., Muratova, N.: A remote sensing based discrimination between climate/human-induced vegetation changes in Central Asia. Management of Environmental Quality: an International Journal, Vol. 19, Issue 5: 579-596, 2008b.

Reuters Exclusive: Peabody, China and Russia teams chosen in mine bid www.reuters.com/article/2011/07/04/us-mongolia-tavantolgoi. 2011.

Sternberg, T., Middleton, N., Thomas, D.: Pressuriced pastoralism in South Gobi Mongolia: what is the role of drought? Transactions of the Institute of British Geographers, 34: 364-377, 2009.

Ulambayar, T., Fernández-Giménez, M.: Following the Footsteps of the Mongol Queens: Why Mongolian Pastoral Women Should Be Empowered. Rangelands, 35: 29-35, 2013.

United Nations Environment Programme (UNEP). Mongolia Faces Critical Water Shortfall Warns UNEP Report. 2011

www.unep.org/roap/Portals/96/Documents/MongoliaWaterReport2011.pdf

Zhang, Y., Munkhtsetseg, E., Kadota T., Ohata, T.: An observational study of ecohydrology of a sparse grassland at the edge of the Eurasian cryosphere in Mongolia. Journal of Geophysical Research, 110, D14103. 2005.

Evaluation of groundwater resources in the upper Tuul River basin, Mongolia

Enkhbayar Dandar[1], Jesús Carrera[1], Buyankhishig Nemer[2]

[1] Institute of Environmental Assessment and Water Research (IDAEA), CSIC, Barcelona, Spain

[2] Department of Geology and Hydrogeology, School of Geology and Mining Engineering, MUST, Ulaanbaatar, Mongolia

Abstract

Aquifers are an important source of water for Mongolia, especially when surface waters are frozen in winter time. Therefore, evaluating recharge is a necessary step for any water planning. Unfortunately, few studies are devoted to groundwater recharge estimation in Mongolia. The objective of this study is to estimate the groundwater recharge using the soil water balance method in order to evaluate groundwater resource from easily accessible climate and soil data in the upper Tuul River basin of northern Mongolia, as a preliminary step for the assessment of groundwater resource. To this end, we use meteorological data from both local weather station and WATCH project and soil data from regional map. We compute an average yearly recharge of, at most, 17 mm/year for the 1993-2001 periods, which was relatively wet. This value is far too small when compared to river discharge (some 80 mm/year) which results basically from groundwater discharge. We conjecture that this underestimation is caused, first, by significant subestimation of rainfall (weather stations are located at low elevations where rainfall is much lower than at high elevations) and, second, to ungauged winter rainfall and permafrost thawing.

Keywords – recharge, potential evapotranspiration, field capacity, discharge

INTRODUCTION

Mongolia has limited freshwater resources distributed unevenly throughout the country. Groundwater is an important source of water supply, especially during winter when surface waters are freeze. In order to evaluate the groundwater resource, it is essential to quantify areal recharge. There are several detailed reviews of methods for estimating recharge (Lerner et al., 1990; Scanlon et al., 2002; Maliva et al., 2012). Recharge can be calculated by soil water balance (Kumar, 2003; Rushton et al., 2006), water table fluctuation (Healy and Cook, 2002), lysimeters (Lerner et al., 1990), natural tracers (chloride mass balance, isotopes etc.) and tracer experiments (Custodio, 2010; Allison et al., 1994) and numerical modeling (Sanford, 2002). These methods have proven useful, but most of them require a lot of fieldwork and expensive to obtain data.

The climate of Mongolia is semi-arid to arid (Batimaa et al., 2007), so that evapotranspiration (ET) is close to rainfall (R). In these cases, small errors in the evaluation of R or ET will cause large relative error in recharge estimates (Gee and Hillel, 1988 and Lerner et al., 1990) and the estimation of recharge with limited data remains one of the most challenging issues in water resources research.

Few studies are focused on recharge estimation in Mongolia. Tserenjav (1981) classified the territory of Mongolia into four recharge zones depending on surface morphology (Batsukh et al., 2006): (1) mountain area with high recharge (31.54-63.07 mm/year), (2) mountain area with medium recharge (3.15-31.54 mm/year), (3) lowland-hilly steppe region with insufficient recharge (1.26-3.15 mm/year), (4) Gobi steppe-hilly region with scarce recharge (less than 1.26 mm/year). Hiller and Jadambaa (2006) estimated the recharge in the 14 sub-basins of the Selenge River basin using the equation of Kudelin.B (1960) which is based on the catchment water balance method. They reported recharge of approximately 6% or 20 mm/year of the average annual rainfall for the Selenge River basin and about 5.3 % or 18.6 mm/year for the entire Tuul River basin. Tadashi et al. (2004) and Tsujimura et al. (2007) evaluated recharge in the Kherlen River basin, eastern of Mongolia by the stable isotope ratios and continuous monitoring of soil water contents. They noted that groundwater recharge can occur only during large precipitation events of more than 30 mm of total rainfall.

Buyankhishig et al. (2007) calibrated a 2-D groundwater model to obtain a set of recharge rate and hydraulic conductivities in alluvium-unconfined aquifer which is located in the central source of water supply for Ulaanbaatar. They found that the source of water to the alluvial aquifer is the Tuul River. In addition, a recent study of the interaction between groundwater in the same alluvial aquifer and river water using a multi-tracer approach presented that the groundwater is mainly coming from the Tuul River (Tsujimura et al., 2013). Still, the Tuul River is fed by groundwater discharge upstream because surface run-off is negligible. Therefore, areal recharge needs to be evaluated throughout the basin in order to properly understand basin hydrology, to evaluate available resources and to assess its response to climate change. To this end, the soil water balance approach is the only method that can be applied with the available data to estimate areal recharge.

The objective of this study is to estimate areal recharge using the soil water balance method in order to evaluate groundwater resources from easily accessible climate and soil data in the upper Tuul River basin, as a preliminary step to the assessment of groundwater resources.

Study area

The study area is located in the Upper Tuul River basin in northern of Mongolia. It includes Ulaanbaatar with a population around 1.37 million (NSO, 2013). Topography in the basin ranges from 1292 to 2773 m above mean sea level and covers an area of 7253 km^2. The Tuul River, which is a tributary of the Selenge River, originates in the Khentii Mountains and flows through Ulaanbaatar, generally from northeast to southwest in a meandering channel. The rivers are completely frozen from December to February (Buyankhishig et al., 2007). Flow resumes in March and discharge gradually increases to peak in the rainy seasons from July and August. The average monthly temperature in the three weather stations of Figure 1 varies from a minimum of -22.6°C in January to a maximum of 16.8°C in August. Precipitation in the selected areas varies significantly from year to year, between 163.1 and 514.1 mm/year. Nearly 80% of the annual precipitation falls between June and September on average. During winter the ground is frozen and precipitation accumulates as snow. Snow usually falls between mid-October and mid-April, and thick snow covers the surrounding mountains which remain covered until early April.

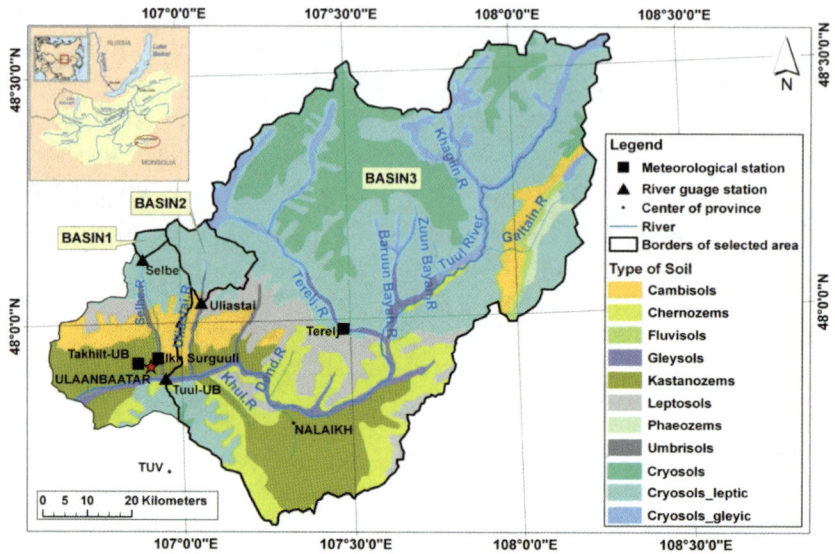

Figure1. Study area and soils of Upper Tuul Basin (National Soil Atlas of Mongolia, 1981)

MATERIALS AND METHODS

The meteorological datasets used in this study were obtained from two sources: the Institute of Meteorology and Hydrology of Mongolia (IMH) for the period of 1993-2013 and the WATCH forcing data (WFD) for period of 1958-2001. The WFD is a half degree resolution daily data set that contains data of eleven variables: rainfall, maximum and minimum temperature, wind speed at height of 10 m, shortwave and longwave radiation and specific humidity (Weedon et al, 2011). For IMH data, we used daily precipitation, maximum and minimum daily temperature, daily wind speed at height of 2 m from 3 meteoro-logical stations. Figure 2 is shown evolution of IMH and WFD meteorological data for the 1993-2001 interval and monthly averages of rainfall and tempera-ture are quite similar for the two datasets (Figure 2a). However, figure 2b, daily values of rainfall are a lot more fluctuating for the IMH data, which reflects that WFD results from spatial averaging.

River discharge used from 3 gauge stations (Selbe-Sanzai, Uliasatai and Tuul-UB) (Figure 1). These stations define the three test basins of this study.

Basin-1 is situated in the upper Selbe River basin. Its area is 33.4 km² and elevation ranges from 1517 to 2026 m.a.s.l. Basin-2 is Uliastai River, a tributary of Tuul River. Its area is about 202.4 km² and elevation ranges from 1484 to 2194 m.a.s.l. Basin-3 is the upper Tuul River basin and its area of 6395.3 km². The elevation ranges from 1292 to 2773 m.a.s.l. Basin-3 includes Ulaanbaatar.

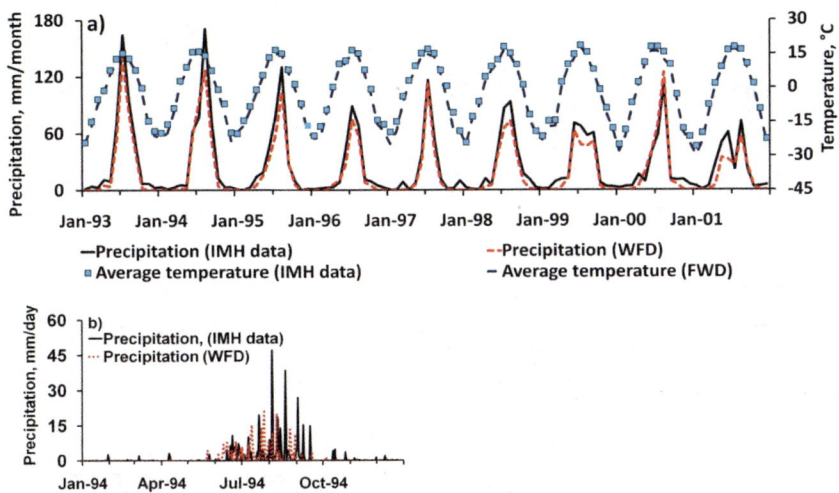

Figure 2. a) Monthly evolution of IMH and WFD meteorological data for the 1993-2001. b) Daily precipitation for IMH and WFD in 1994

Soil properties

As soil data, we used the soil map created by Mongolian and Russian research-ers using aerophoto and satellite (National Soil Atlas of Mongolia, 1981). Soils are classified by reference soil group (Working Group World Reference Base, 2006) (Figure 1). The majority of the study area (58.8%) is covered by cryosols (including cryosols with leptic and cryosols with gleyic), 12.8% by kastano-zems, 9.5% by chernozems, 7.3% by leptosols, 6.4% by cambisols, 2.9% by gleysols and 0.7% by every chernozems, phaeozems and umbrisols. For each soil type, field capacity and wilting point were obtained from from Schroeder et al. (1994) and soil thickness and texture from FAO (2001a) (Table 1).

Soil name	Area, km^2	Thickness, cm (*)	The soil texture(*)	F.C, vol/vol (**)	W.P, vol/vol (**)
Cambisols	465.6	70	C-L	0.373	0.266
Chernozems	688.8	70	Si - C –L	0.393	0.277
Fluvisols	54.2	50	C	0.378	0.251
Gleysols	215.0	50	S - C	0.366	0.288
Kastanozems	933.1	50	C	0.419	0.332
Leptosols	527.6	25	S - C -L	0.305	0.202
Phaeozems	51.0	60	Si – L	0.360	0.203
Umbrisols	53.2	50	Si - C –L	0.342	0.210
Cryosols	1005.7				
Cryosols_leptic	2826.0	100	S-L	0.190	0.085
Cryosols_gleyic	432.9				

Abbreviation: F.C-Field Capacity; W.P- Wilting Point; S-Sand; C-Clay; L-Loam; Si-Silt

*   - *source:*FAO. 2001a. Lecture notes on the major soils of the world

**- *source:* Schroeder et al., 1994

Table 1. The soil properties

Soil water balance method for estimating the groundwater recharge

Areal recharge is the amount of rainfall that infiltrates below the root zone. It can be described by the soil water balance (Carrera et al., 2002) as follows:

$$\frac{dV_t}{dt} = P_t - ES_t - ET_t - R_t \tag{1}$$

where V_t is the volume of water in the soil (root zone), P_t is the rainfall, ES_t is the surface runoff, ET_t is the potential evapotranspiration, and R_t is the recharge. All terms are expressed as volume of water per unit area and time (mm per day or mm per month). Rainfall is the primary source of water to the soil water cycle. Some part of the rainfall may be reduced due to surface runoff or interception. We have not considered these terms because of small vegetation cover, because we are interested in long term recharge and because little sur-face runoff occurs in the Tuul Basin.

The most important term for the soil water balance method in semi-arid regions is ET, which is obtained from the potential evapotranspiration (PET). PET is defined as the amount of water that evaporates and transpires from a vegetated surface with no restrictions and both processes depending on solar radiation, air temperature, relative humidity and wind speed. There are several empirical methods to estimate PET (Lu et al., 2005; Zotarelli et al., 2009). We chose the Penman-Monteith (PET-PM) method as described by Allen et al (1998), which was used as the standard for comparison and used daily aver-ages of the WFD and IMH data:

$$PET_{FAO} = \frac{0.408 \, \Delta \, (Rn - G) + \gamma \left(\frac{900}{T + 273}\right) u_2(e_s - e_a)}{\Delta + \gamma(1 + 0.34 \, u_2)} \tag{2}$$

where PET_{FAO} is the potential evapotranspiration (the reference crop evapo-transpiration of FAO, in mm/d), Rn is the net radiation ($MJ/m^2/day$), G is soil heat flux density ($MJ/m^2/day$), T is mean daily air temperature ($^\circ C$), u_2 is wind speed at 2 m height (m/s), e_s is saturation vapour pressure (kPa), e_a is actual vapour pressure, Δ is the slope of the vapour pressure curve (kPa/$^\circ$C), and γ is the psychometric constant (kPa/$^\circ$C).

The calculation procedure for the PET-PM method requires accurate measurements of air temperature and relative humidity, solar radiation and wind speed. Relative humidity and solar radiation are not available in this study area. Thus, e_a was obtained, assuming that the minimum temperature is close to dewpoint (T_{dew}) (Zotarelli et al., 2009). The solar radiation was derived from the extraterrestrial radiation and relative sunshine duration (Allen et al, 1998).

Additionally, for PET, we selected the Hargreaves and Samani (1985) method (PET-Harg). It requires only maximum and minimum air temperature, available at most weather stations and extraterrestrial radiation (Ra). The method is derived by assuming that cloud cover can be approximated by (Tmax-Tmin) and reads (also in mm/day):

$$PET_{Harg} = 0.0023 \, Ra \, (T + 17.8)\sqrt{T_{max} - T_{min}} \qquad (3)$$

where Tmax, Tmax, T are the maximum, minimum and average daily air temperatures (°C), respectively. Ra was calculated from the station latitude using the equation recommended by the FAO (Allen et al, 1998).

Groundwater recharge (R) occurs when the soil water content exceeds the field capacity. It is calculated using equation (1) for daily time steps from 1993 to 2013 for IMH data and from 1958 to 2001 for the WFD. Mean areal recharge for the whole basin can be written as:

$$R = \frac{\sum_{i=1}^{n}(A_i R_i)}{\sum A_i} \qquad (4)$$

where i is the soil type number, Ai is its area and Ri is its areal recharge, estimated using the above methodology. When the using IMH data, Ri was computed by averaging the result obtained with data from each station separately.

Result and discussion

Potential Evapotranspiration (PET)

Daily potential evapotranspiration (PET) estimated from 1993 to 2013 using IMH data and both PET-PM and PET-Harg methods, from 1958 to 2001 using the WFD and the PET-PM method. Monthly integrated and averaged results are shown in figure 3 for 1993-2001 when data are available for all methods. PET-Harg (IMH data) had the highest value while PET-PM (IMH data) and PET-PM (WFD) were approximately equal. The highest PET occurs between May and July, with a maximum value in June: 102.13 mm for PET-PM (IMH data), 102.45 mm for PET-PM (WFD) and 160.02 mm in June for PET-Harg (IMH data).

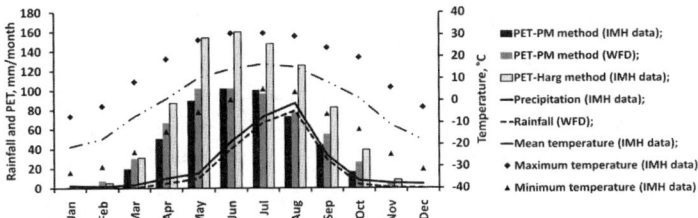

Figure 3.The monthly comparative of potential evapotranspiration (PET), precipitation and temperature for 1993-2001

Total yearly average PET with PET-PM (IMH) method estimated 504.15 mm/year, 534.84 mm/year for PET-PM (WFD) and 878.88 mm/year for PET-PM (IMH) in 1993-2001. Figure 3 shows the maximum, minimum and mean temperature for IMH data and comparison of the precipitation for WFD and IMH data. The PET-PM method using WFD and IMH data yield very similar results, but very different to those obtained with the PET-Harg method which has worked well for Mongolian grassland area (Tuya et al., 2006).

Groundwater recharge (R)

We utilized the three different PET calculations above for estimating the groundwater recharge in the upper Tuul River basin. The resulting monthly recharge is shown in figure 4 and the evolution of annual recharge in figure 5. Recharge occurs in the rainy season from June to September.

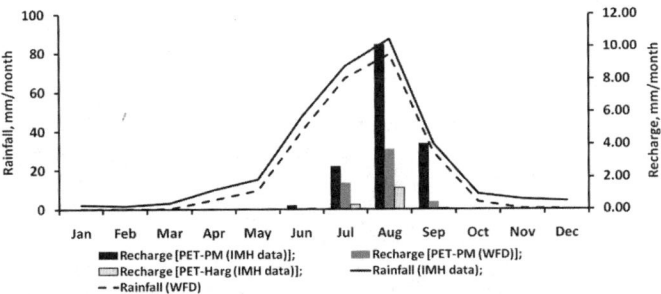

Figure 4. Monthly recharge for 1993-2001 and monthly precipitation for IMH data and WFD

In figure 4, the highest recharge occurs in August and it equals 10.3 mm for the PET-PM (IMH data), 3.69 mm for the PET-PM (WFD) and 1.32 mm for using the PET-Harg (IMH data). The recharge from PET-PM (IMH data) is much greater than the recharge from PET-PM (WFD), both because rainfall is somewhat higher in IMH data than in WFD and, especially, because IMH data contains much larger high rainfall events than WFD. These events are ones that cause most recharge. The recharge obtained with PET-Harg is lower than the recharge with PET-PM, because the PET-Harg is higher than the PET-PM. The mean annual recharge using PET-PM (IMH data) is 8.58 mm/year (or 3.2 % of rainfall) for 1993-2013, the recharge using PET-PM (WFD) is 5.4 mm/year for 1958-2001 and the recharge using PET-Harg (IMH data) is 1.09 mm/year for 1993-2013. For 1993-2001, which are wet years, the mean annual recharge that using PET-PM (IMH data) is of 17.02 mm/year, the recharge using PET-PM (WFD) is 5.72 mm/year and the recharge using PET-Harg (IMH data) is 1.72 mm/year (Figure 5).

Figure 5. Evolution of annual recharge and rainfall

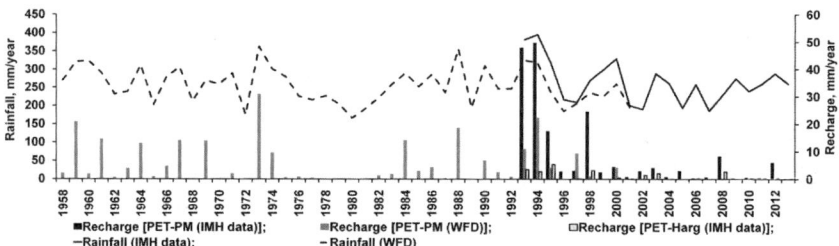

River discharge (Q) for selected basins

As a water to verify the approach, we compare computed recharge with river discharge at the three basins of Figure 1. The comparison implicitly assumes that there is no surface runoff which is consistent with observations. This computation also acknowledges that recharge depends on soil moisture condition (field capacity and wilting point), root zone and the soil type. Computed and observed discharges are shown figure 6. Calculated river discharge using actual rainfall data (red line in figure 6) is much lower than observed for three rivers. We repeated the calculation multiplying rainfall by 1.5 (green line) and by 2.5 (violet line). It is clear that the rainfall data we have used cannot explain the observed river discharge. This can be attributed to several factors. First rainfall is likely to increase with elevation. Yet, all meteorological stations are located in the low area, mostly not in mountain areas. Second snowmelt process and permafrost phenomena can by influenced to estimation of the groundwater recharge.

CONCLUSION AND DISCUSSION

Groundwater recharge was calculated in the upper Tuul River basin by the soil water balance method using daily meteorological data from IMH for the period of 1993-2013 and the WATCH forcing data (WFD) for the period of 1958-2001.

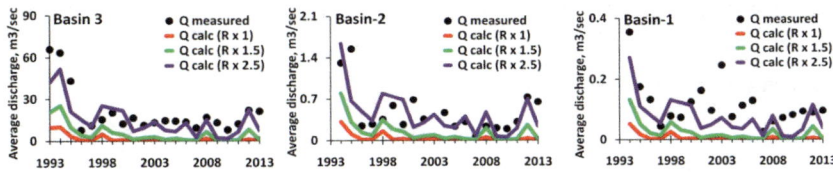

Figure 6. Calculated and observed yearly discharge for the three basins from 1993 to 2013. Notice that comparable values can only be obtained by multiplying rainfall by a factor of 2.5.

Potential evapotranspiration (PET) was calculated by two methods (Penman-Monteith and Hargreaves-Samani). PET computed with the better method was much higher than with the PM method. For examples, PET is highest between May and July, with maximum values in June of 102.13 mm for PET-PM (IMH data), of 102.45 mm for PET-PM (WFD) and of 160.02 mm for PET-Harg (IMH data).

We used these three PET time series for estimating the groundwater recharge in the upper Tuul River basin. The highest monthly recharge was computed in August when rainfall is also highest. The recharge obtained PET-PM (IMH data) is much greater than the recharge used PET-PM (WFD). The mean annual recharge in the study area during the period 1993-2001 obtained with the PET-PM (IMH data) is of 17.02 mm, 5.72 mm for PET-PM (WFD) and 1.72 mm for PET-Harg (IMH data). These values are much smaller than river discharge (some 80 mm/year). Since river flow is essentially groundwater discharge, we conclude that the recharge values we have computed are too small.

We conjecture that recharge underestimation can be attributed to several factors. First, the most important actual rainfall may be much larger than measured because rainfall increase with elevation and the three weather stations are located at low altitude. Second, measured rainfall is zero during winter. In reality, it is not, but falls as snow in the mountain. Third, moisture can be captured by the permafrost. We are currently working to account for these factors.

Acknowledgements

We thank the project of IWRM in Water Authority of Mongolia (old name) and the Institute of Environmental Assessment and Water Research (CSIC-IDAEA) of Spain for supporting the first author. Help was also provided by Gonzalo Sapriza, Enric Vázquez-Suñe and D. Tuvshinjargal.

References

Allen R.G., Pereira L.S., Raes, D. and Smith, M. (1998). *Crop evapotranspiration: guidelines for computing crop water requirements.* FAO Irrigation and Drainage Paper, No.56, Rome, Italy, 328 p

Allison G. B, Gee G.W, Tyler S.W. (1994). *Vadose-zone techniques for estimating groundwater recharge in arid and semiarid regions.* Soil Science Society of America Journal, 58(1), 6-14.

Batimaa P., Natsagdorj L., Batnasan N. (2007). *Vulnerability* of *Mongolia'sPastoralists* to *ClimateExtremes* and *Changes. Climate change and Vulnerability. Edited by Neil Leary, Cecilia Cande, Jaoti Kulkarni, Anthany Nyong and Juan Pulhin. London. pp 67-87.*

Batsukh N., Buyankhishig N, Oyun D, Boldbaatar YA. (2006). *Sustainable development and groundwater of Mongolia.* 34th congress of International Association of Hydrogeologists, Beijing, China.

Buyankhishig N., Masumoto K, Aley M. (2007). *Parameter estimation of an unconfined aquifer ofthe Tuul river basin Mongolia.* Proc.52nd International Scientific Colloquium. Universitätsverlag Ilmenau. Vol. II. pp. 67-72 S. Röll, S.

Carrera J. R., Saaltink M.W. (2002). *Balance de agua en el suelo. Métodos hidrometeorológicos.* ETSI Caminos. UPC (in spanish) [Carrera J.R.,

Saaltink M.W. (2002). Soil water balance. Hydrometeorological methods. ETSI Caminos. UPC]

Custodio E, (2010). *Estimation of aquifer recharge by means of atmospheric chloride deposition balance in the soil.* Contributions to science 6 (1): 81-97

FAO. (2001a). *Lecture notes on the major soils of the world (with CD-ROM),* by P. Driessen, J. Deckers, O. Spaargaren & F, Nachtergaele, eds. World Soil Resources Report No. 94. Rome.

Hargreaves G. H. and Samani, Z., (1985). *Reference Crop Evapotranspiration from Temperature, Appl.* Eng. Agric., vol. 1, no.2, pp. 96-99.

Healy R.W, Cook P.G. (2002). *Using groundwater levels to estimate recharge.* Hydrogeology journal, 10(1), 91-109.

Hiller B.T. and Jadambaa N. (2006). *Groundwater use in the Selenge River basin, Mongolia.* The full papers of the 34th congress of International Association of Hydrogeologists, Beijing, China.

Kumar C.P. (2003). *Estimation of ground water recharge using soil moisture balance approach.* Journal of soil and water conservation, Soil Conservation Society of India, 2(1-2), 53-58.

Lerner D.N, Issar A.S, Simmers I.1990. *Groundwater recharge, a guide to understanding and estimating natural recharge.* International Association of Hydrogeologists, Kenilworth, Rep 8, 345 pp

Lu, J. B., Sun, G., Mcnulty, S. G. and Amatya, D. M. 2005. *"A comparison of six potential evapotranspiration methods for regional use in the southeastern United States."* J. Am. Water Resour. Assoc., 41(3), 621–633.

Maliva R, Missimer T. (2012). *Recharge Measurement in Arid and Semiarid Regions. Arid Lands Water Evaluation and Management, Environmental Science and Engineering.* DOI: 0.1007/978-3-642-29104-3 11. (pp. 247-276). Springer Berlin Heidelberg.

NATIONAL SOIL ATLAS OF MONGOLIA, (1981) (Scale 1:1000 000)

NSO, (2013) *National Statistical Office of Mongolia, national report of 2013 population*. Ulaanbaatar. Mongolia

Rushton K.R, Eilers V.H. M, Carter R.C. (2006). *Improved soil moisture balance methodology for recharge estimation*. Journal of Hydrology, 318(1), 379-399.

Sanford, W. (2002). *Recharge and groundwater models: an overview*. Hydrogeology Journal, 10(1), 110-120.

Scanlon B. R, Healy R.W, Cook P.G. (2002). *Choosing appropriate techniques for quantifying groundwater recharge*. Hydrogeology Journal, 10(1), 18-39.

Shroeder P.R, Dozier T.S, Zappi P.A, Mcenroe B.M, Sjostrom J.W, Peyton R.L. (1994). *The hydrologic evaluation of landfill performance (HELP) model: engineering documentation for version 3*. EPA/600/9-94/168b, U.U. Environmental Protection Agency Risk Reduction Engineering Laboratory. Cincinnati, Ohio.

Tanaka T., Abe Y., Tsujimura M. (2004). *Groundwater recharge process in the Kherlen river basin*. Proc. 3rd Int. Worksop on Terrest. Change in Mongolia, 15-18.

Tsurjimura M., Abe Y., Tanaka T., Shimada J., Higuchi S., Yamanaka T., DAVAA G, OYUNBAATAR D. (2007). *Stable isotopic and geochemical characteristics of groundwater in Kherlen River basin, a semi-arid region in eastern Mongolia*. Journal of Hydrology 333:47–57

Tsurjimura M., Ikeda Y., Tanaka T., Janchivdorj L., Erdenechimeg B., Unurjargal D., Jayakumar R. (2013). *Groundwater and surface water interactions in an alluvial plain, Tuul River Basin, Ulaanbaatar, Mongolia*. Sciences in Cold and Arid Regions, 5(1): 0126-0132

Tuya S., Batbayar J., Kajiwara K. and Honda Y. (2006). *A comparison of five potential evapotranspiration methods and relationship to NDVI for re-*

gional use in the Mongolian grassland. International Archives of the Photogrammetry, Remote Sensing and Spatial Information Science, Volume XXXVI, Part 6, Tokyo Japan

Weedon G. P., Gomes S., Viterbop, Shuttleworth W. J., Blyth E., Osterle, H. (2011). *Creation of the WATCH Forcing Data and Its Use to Assess Global and Regional Reference Crop Evaporation over Land during the Twentieth Century.* J. Hydrometeor, 12(5), 823-848. doi:10.1175/2011JHM1369.1

Working Group World Reference Base International Union of Soil Sciences. (2006). World reference base for soil resources 2006: A framework for international classification, correlation and communication. FAO.

Zotarelli L., Dukes M. D., Romero C. C., Migliaccio K. W. and Kelly M. T. (2009). Step by Step Calculation of the Penman-Monteith Evapotranspiration (FAO-56) Method. University of Florida AE459.

II. Environmental pollution impacts of anthropogenic activities

Influence of urban settlement and mining activities on surface water quality in northern Mongolia

Gunsmaa Batbayar[1,2,3], Daniel Karthe[1,2], Martin Pfeiffer[3], Wolf von Tümpling[1], Martin Kappas[2]

[1] Department Aquatic Ecosystem Analysis and Management, Helmholtz Centre for Environmental Research, Magdeburg, Germany
[2] Institute of Geography, Georg-August University, Göttingen, Germany
[3] School of Arts and Sciences, National University of Mongolia

Abstract

Until today information about water quality in Northern Mongolia is rare. For this reason the aim of this snap short study was to assess the current chemical status in the region based on 40 surface water samples collected between May and July in 2014. As main outcome of this study we showed that chemical water quality varied in rivers, which is potentially due to geological differences, anthropogenic pollution and dilution/concentration effects related to evaporation and precipitation. The study indicates that chemical water quality is impacted in particular by mining, industry and urban wastewater discharge. However, a more in depth and long-term study is needed in order to assess the spatial and temporal contamination pattern and evaluate their ecological and public health relevance.

Introduction

Mongolia receives very limited precipitation with an annual mean value of round about 400 mm in the northern area. Some 70 to 90 % of the precipitated water evaporates back into the atmosphere (Davaa et al. 2006, Battogtokh et al. 2012). The study area, which covers the Mongolian part of the Selenga River basin, including the three largest Mongolian cities Ulaanbaatar (UB) (ca. 850,000 inhabitants), Erdenet (ca. 80,000 inhabitants) and Darkhan (ca. 70,000 inhabitants), is characterized by a highly continental climate with very cold winters and a limited natural water availability (Minderlein & Menzel 2015,

Karthe et al. 2014, Karthe et al 2015, MEGD 2012). Within this region, a considerable part of the mining and industrial activities and more than half of Mongolia's population with in total more than 3 Mio inhabitants are concentrated (Karthe et al. 2014). In recent years growing urbanization and mining industry have polluted surface and underground water resources significantly (Hoffmann et al. 2010, Pfeiffer et al. 2015, Thorslund et al. 2012). This has also impacted the associated ecosystems (Avlyush 2011, Krätz et al. 2010). Until now almost half of the Mongolian people receive water from sources regarded as unsafe, such as unprotected wells, rivers, run-off or snow (Javzan et al. 2013). Investigations conducted by the Institute of Geo-ecology of the Mongolian Academy of Sciences on drinking water wells in Ulaanbaatar (Javzan et al. 2013) have shown that some wells are biogenic/organic polluted and have elevated concentrations of trace elements, like As (up to 13 µg/l), Se (up to 20 µg/l), Al (up to 0.68 mg/l), Sr (up to 2.6 mg/l) and Mo (up to 0.35 mg/l). Moreover elevated levels of U (up to 57 µg/l) have been reported (Nriagu et al. 2012). While extensive water quality monitoring in Mongolia is still in its beginning, elevated levels of nutrients, arsenic and uranium have recently been documented in surface water, groundwater, soils/sediments and urban vegetation for several locations in northern Mongolia (Hoffmann et al. 2015, Pfeiffer et al. 2015, Nriagu et al. 2012).

Material and methods

A sampling campaign was carried out in the Kharaa, Orkhon and Selenga River Basins between 27 May and 27 July 2014. Totally 40 water samples were collected from the study area (Figure 1). On site we collected the main parameters water temperature, pH and electrical conductivity using a portable multimeter (WTW, Multi 3320). To determine the dissolved concentration of the investigated parameters, the water samples were filtered through Minisart cellulose acetate filters. Afterwards all water samples were filled into brown glass and Sarstedt tubes, bubble-free. At the laboratories the filtered samples were stored at 4 ^{0}C in a refrigerator. For element analysis the samples were preserved with nitric acid (HNO$_3$; pH < 2). All investigated elements were determined using ICP-MS, ICP-OES or CV-AAS. ICS was used to measure Cl$^-$ and SO$_4^{2-}$ concentrations. Organic carbon was determined as total concentration

(TOC) and after filtration as dissolved fraction (DOC) using a carbon analyzer. Total nitrogen, ammonia, nitrate, nitrite and soluble phosphorus were determined using continuous flow analysis. For total P and total N, we stored water samples in 30 ml HDPE bottles and preserved them by 350 μl H_2SO_4 (1:4) before analysis (Table1).

	Country	Laboratory	Methodology	Parameters
	Germany	UFZ Helmholtz center for environmental research laboratory	ICP-MS	Ba, Be, Li, Rb, Sr, B, Al, As, Cd, Co, Cu, Cr, Pb, Sn, Bi, Mo, Sb, Ag, Tl, Ti, V, U
			ICP-OES	Ca2+, K+, Mg2+ , Na+, Fe, Mn, Ni, Zn
			ICS-3000	Cl-, SO4 2-
Water analysis			Carbon analyzer	TOC,DOC
			Continuous flow analysis	Total P, Total N
			Atomic fluorescence	Hg
	Mongolia	Central Geological laboratory	ICP-MS	As, U, Mn

Table 1. Water analysis methodology and standards.

Result and Discussion

Here we primarily focus on comparisons of our samples with the Mongolian water quality general requirement (MNS 4586:1998), and Water Framework Directive (WFD) (Figure 3). In some cases the Mongolian quality general requirement is stricter than the WFD and even the EPA guideline (Figure 3). In total 40 random water samples were collected. The parameters NO3-N, TNb, SRP, DOC, TOC, B, Cl,SO42-, Ca+, K, Mg2+, Na+, U, Ba, Be, Li, Rb, Sr, Cd, Co, Cr, Sn, Bi, Tl, Ti, and V did not exceed the MNS 4586:1998 and WFD guidelines[1].

---

[1] Note: the underlined parameters are not set in the guideline.

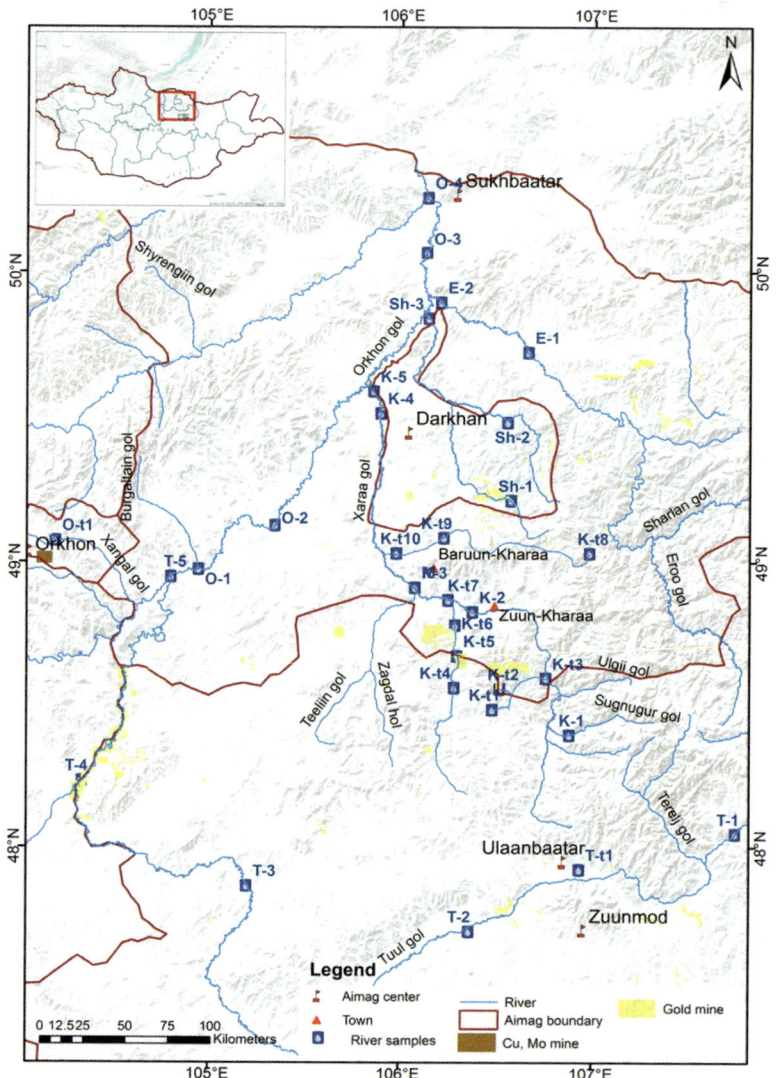

Figure 1: Map of the study area, indicating sampling locations and mining areas.

Parameters	Unit	WFD	MNS 4586:1998	Mean	SD	Min	Median	Max
Ammonium-N	mg/l	0.5	0.5	0.16427	0.49812	0.011	0.039	2.8
NO_2-N	mg/l	0.5	0.02	0.00776	0.00558	0.006	0.006	0.033
NO_3-N	mg/l	50	9	0.43021	0.69863	0.024	0.104	3.03
Total-N	mg/l			1.42488	0.91515	0.387	1.24	4.7
Total-Phosphate-P	mg/l	0.1	0.1	0.15691	0.13434	0.01	0.126	0.458
SRP	mg/l			0.02	0.0414	0.003	0.007	0.235
DOC	mg/l			6.87	3.00814	3.81	5.635	15.2
TOC	mg/l			8.69515	5.09023	3.35	7.32	28.3
Boron	µg/l	1000		29	23.2002	10	18	98
Cl^-	mg/l	250	300	6.09212	4.67797	1	5.58	20.8
SO_4^{2-}	mg/l	250	100	25.50121	55.9311	2.66	14.3	332
Ca_2^+	mg/l			33.29939	21.6772	4.3	32.1	131
K^+	mg/l			3.02736	1.52248	0.603	2.92	6.81
Mg^{2+}	mg/l			10.61103	8.15586	0.834	10	45.1
Na^+	mg/l	200		13.21697	12.4708	1.09	10.5	55.9
Uranium	µg/l			6.52727	5.83632	0.5	3.3	20.8
Barium	µg/l			59.42424	39.9765	11	48	172
Beryllium	µg/l			1	0	1	1	1
Lithium	µg/l			10.48485	6.29544	1	10	32
Rubidium	µg/l			8.63939	9.52585	0.5	5.1	38.6
Strontium	µg/l			281.27273	264.634	30	245	1610
Aluminum	mg/l	0.2		4.85085	5.50284	0.236	2.33	21.1
Arsenic	µg/l	10	10	12.47576	41.6382	0.5	5	243
Cadmium	µg/l	5	5	0.2	0	0.2	0.2	0.2
Cobalt	µg/l		10	2.07576	2.13834	0.4	1.1	8.1
Copper	µg/l	2000	10	16.95455	54.0641	2	5.1	316
Chromium	µg/l	50	50	7.12121	7.20485	0.6	4.5	29.3
Fe	mg/l	0.2		4.68909	5.45376	0.23	2.16	19.7
Mercury	µg/l	1	0.1	0.01469	0.01292	0.004	0.009	0.067
Manganese	mg/l	0.05	0.1	0.13383	0.11195	0.007	0.0925	0.418
Nickel	µg/l	20	10	6.02121	5.29308	0.9	4	20.5

Lead	µg/l	10	10	2.55758	2.34295	0.5	1.5	7.8
Tin (Sn)	µg/l			1	0	1	1	1
Zink	mg/l		10	0.02478	0.01838	0.009	0.0167	0.0617
Bismuth	µg/l			0.80909	0.03844	0.8	0.8	1
Molybdenum	µg/l		250	7.76061	23.5752	0.7	3.3	138
Antimony (Sb)	µg/l	5		0.35758	0.17145	0.3	0.3	1.2
Silber	µg/l			0.97576	0.13926	0.2	1	1
Thallium	µg/l			1	0	1	1	1
Titan	µg/l			217.33333	258.644	9	91	1010
Vanadium	µg/l			12.55455	11.8528	0.6	8.6	46.5

Table 2: Overview of analytical results as compared to Mongolian and international surface water standards.

Compared to the Mongolian guideline the following limits were slightly exceeded: NO_2-N (max 1.6fold, n = 2), Nickel (max 2.05fold, n = 1), SO_4^{2-} (max 3.32fold, n = 1), Manganese (max 4.18 fold, n = 16), Total-P (max 4.58fold, n = 23), Ammonium-N (max 5.6fold), Arsenic (max 24.3 fold, n = 9), Copper (max 31.6 fold, n = 10). Following parameters exceeded the WFD guideline: Nickel (max 1.025fold, n = 6), SO_4^{2-} (max1.32 fold), Manganese (max 8.36fold), Iron (max 98.5fold, n = 37), Aluminum (max 106fold, n=33). However, the Copper and NO2-N concentrations met with the WFD guideline. We found elevated Uranium U (n = 9) at Boroo (Measurement points: Kt4-Kt6) and Gatsuurt River (Kt1-Kt3). The Aluminum concentration of all river water samples exceeded the guideline value.

The Orkhon River is 1,124 km long, the longest river of Mongolia (Measurement points: O1-O4). The most polluted tributary of the Orkhon River is the Khangal River (Ot1), the reason for the high pollutant concentrations that we measured are the Erdenet Cupper Molybdenum industry, Erdenet city, as well as low water flux during our study (discharge on site 35 m3/sec). Primary result indicated direct effects from the tailing impoundment and Erdenet city on Khangal River, which is downstream of the Erdenet industry and city. Furthermore high concentration of Copper and Molybdenum as well as nutrients were found in the Khangal River (Ot1). The maximum concentrations of Nickel occurred downstream of the Orkhon River (O3, O4), after the measure sites K5, Sh3, and at E2 at the confluence. This could be explained by our samples down-stream Sharyn River (Sh3) which is influenced by mining activities before it joins with the Orkhon River. Other parameters showed similar trends and heavy metals tended to increase northward (Figure 2).

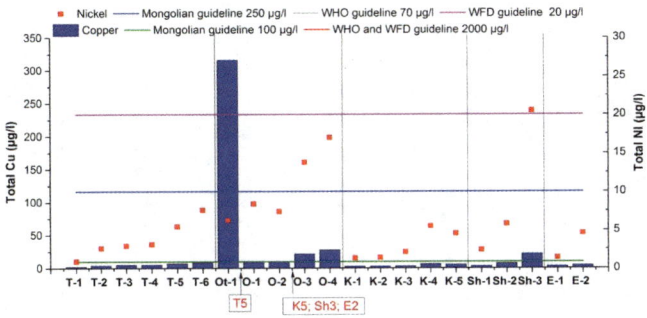

Figure 2. Concentration between Cu and Ni along our measuring track, downstream. T = Tuul, O = Orkhon, K = Kharaa, Sh =Sharyn River, E =Eroo River. Numbers show position downstream as in Fig. 1.

Water quality variables including nutrients are illustrated in Figures 3 to 6. The content of Fe, Ni, Al, U, Cu, Mo, As, Mn, Ca, Mg and SO_4 increased in down-stream direction along each river.

Maximum concentration of nutrient concentrations occurred in Khangal (Ot-1) and (O3) after the K5, Sh3, E2 confluence (Fig. 2 and 3). The concen-trations of TNb and NO_3-N were high downstream of the Sharyn River. We can easily see a nitrification process on O1, O2 and O3 after points Ot1, K5, Sh3 and E2 respectively. Concentration of Ammonium was slightly elevated at K5, Sh3 and E2 and it flows to Orkhon River (O3). The downstream part of the Orkhon Basin stretches from Erdenet to the junction of the Orkhon and Selenga rivers. The concentration of soluble reactive phosphorus (SRP) was slightly elevated on Khangal River (Ot-1), but it decreased along the river.

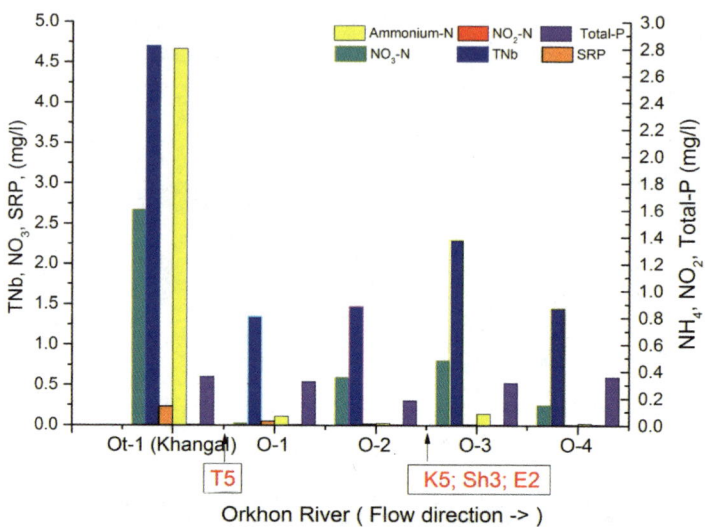

Figure 3: Nutrient concentrations along the Orkhon River.

The Tuul River (T5) is major left tributary of the Orkhon River and it flows through the capital city of Mongolia, Ulaanbaatar (between T1 and T2). We started the water sampling from the Bosgo Bridge (T1) which is located in the upstream section of the Tuul River, which shows (near-) natural reference conditions. The Selbe River is a tributary of the Tuul River which enters the Tuul downstream of Ulaanbaatar. We found high concentration of NO_2-N at Tt1 which indicates the direct influence of untreated waste water from the ger settlement and urban area as well as oxygen depletion (Figure 4). Both maxima may be interpreted as a consequence of anthropogenic eutrophication (nutrient input) and have a direct influence on aquatic life. Water quality continued to change further downstream the Tuul River. The sampling point T2 shows the impacts of a poorly working waste water treatment plant (WWTP) and samples T4 and T5 demonstrate the impacts of mining activities (Zaamar mine area). Location T5 marks the junction of Tuul and Orkhon River (Figure 4).

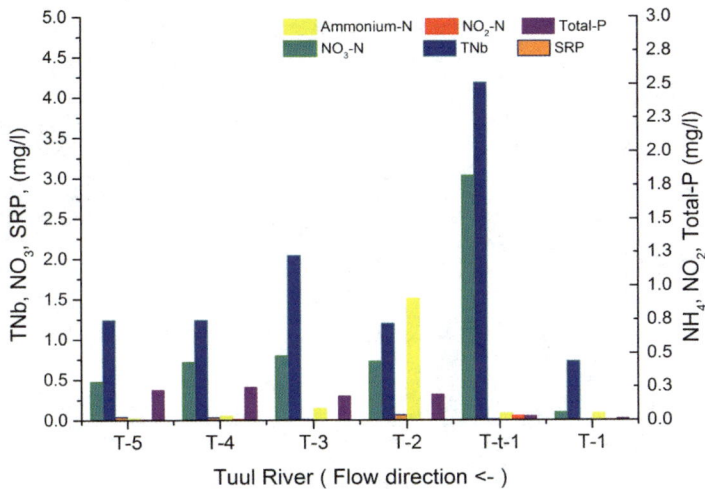

Figure 4: Nutrient concentrations along the Tuul River.

The downstream tributaries of Orkhon River are the Kharaa, Eroo and Shariin Gol River which originate from the Khentii Mountains. These rivers are affected by gold mining activities (Figures 5 and 6).

The area upstream of the Kharaa River (K1) is not affected by human activities. The Kharaa River is one of the main left tributaries of the Orkhon River and extensive studies on its hydrology, water quality and ecological state have been conducted by the MoMo project (Hofmann et al. 2010, Karthe et al, 2015, Menzel et al. 2011) and UNDP-GEF project (Punsalmaa et al. 2013).

In Kharaa river the concentration of Ammonium was slightly elevated after the little town Zuun-Kharaa (K2), Baruun-Kharaa (K3) and Darkhan city (K5), respectively (Fig. 5). Compared to sampling point K1 the following sampling sites had elevated measurements of NO₃-N (K4, K5) and slightly elevated TNb concentration at K3 to K5 (Figure 5).

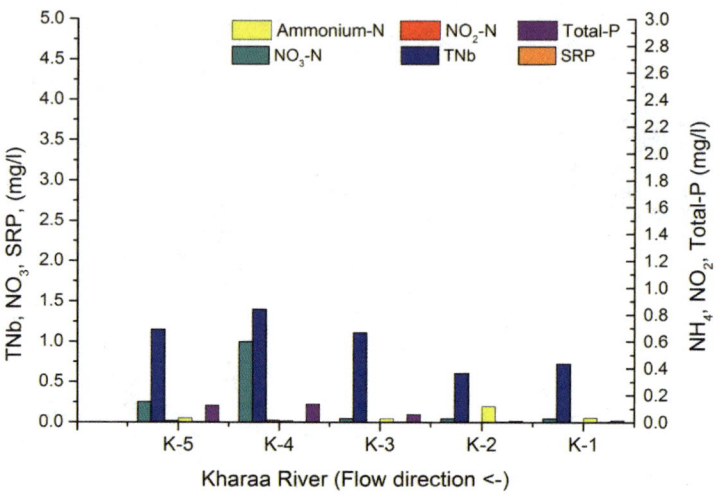

Figure 5: Nutrient concentrations along the Kharaa River.

The Sharyn and Eroo rivers are downstream tributaries of the Orkhon River. The downstream part of the Sharyn River is highly polluted by mine activities compared to the other rivers. Especially the concentrations of TNb, Ammonium-N and Total-P were high (Figure 6).

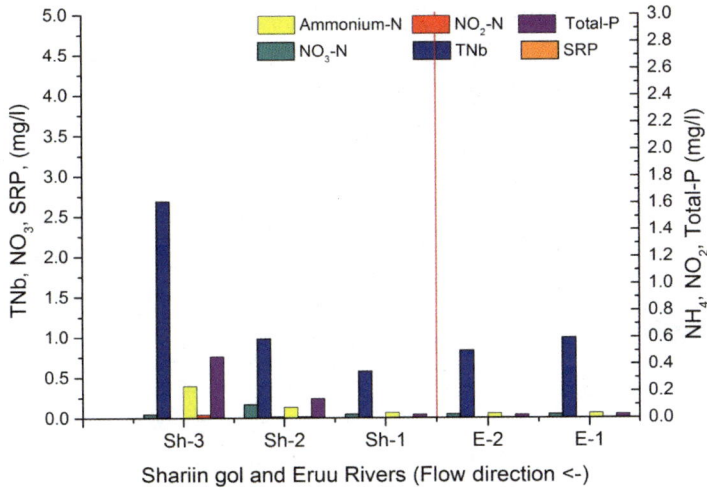

Figure 6: Nutrient concentrations along the Eroo and Shariin River.

Trace elements in Tuul, Kharaa, Eroo and Sharyn Gol River, showed high concentrations and this pollution of its tributaries negatively affected the water quality in the Orkhon River.

Conclusion

The results presented here only provide a snapshot of problems identified during a single sampling campaign. Chemical water quality in Northern Mongolia is impacted by various anthropogenic activities including mining and industry. Uranium was also detected in notable concentrations in several surface water samples from rivers. Moreover, river water samples occasionally exceeded drinking water or surface water guidelines for Al, As, Mn, Cu, Fe and Ni. In the future, an assessment of seasonal changes in water quality due to climatic reasons and variations of anthropogenic impacts from point and non-point sources is necessary.

Acknowledgements

This research was financially supported by the German Academic Exchange Service (DAAD; scholarship number A/12/97034) and the German Federal Ministry of Education and Research (BMBF) in the framework of the FONA (Research for Sustainable Development) initiative (grant no. 033L003).

References

Avlyush, S. (2011): Effects of surface gold mining on macro invertebrate communities. A case study in river systems in the North-East of Mongolia. Saarbrücken: Lambert Academic Publishing.

Battogtokh B., Woo N., Nemer B. (2012): Environmental reconnaissance of the Shivee-Ovoo coalmine area, Mongolia. Environmental Earth Science. 67:1927–1938. DOI:10.1007/s12665-012-1633-z.

Batima P., Natsagdorj L., Gombluudev P., Erdenetsetseg B. (2005): Observed Climate Change in Mongolia. AIACC Working Paper No.12

Davaa G., Oyunbaatar D., Sugita M. (2006): Surface water quality. Environmental Book Mongolia, Vol.2006, pp 55-82

Hofmann J., Venohr M., Behrendt H., Opitz D. (2010): Integrated water resources management in central Asia: nutrient and heavy metal emissions and their relevance for the Kharaa River Basin. Mongolia. Water Science and Technology 62(2): 353-363. DOI:10.2166/wst.2010.262

Hofmann J., Watson V., Sharaw B. (2015): Groundwater quality under stress: contaminants in the Kharaa River Basin (Mongolia). Environmental Earth Science. DOI:10.1007/s12665-014-3148-2

Javzan Ch., Battuya B., Unurjargal D., Enkhtuya M., Udvaltsetseg D., Batkhishig O., Saijaa R. (2013): Impact of Ulaanbaatar city's Ger district pollutant sources into ground water quality. Academy of Science, Institute of Geo-ecology, Project report.

Karthe D., Kasimov N., Chalov S., Shinkareva G., Malsy M., Menzel L., Theuring P., Hartwig M., Schweitzer C., Hofmann J., Priess J., Lychagin M. (2014): Integrating Multi-Scale Data for the Assessment of Water Availability and Quality in the Kharaa-Orkhon-Selenge River System. Geography, Environment, Sustainability 3(7): 65-86

Karthe D., Chalov S., Borchardt D. (2015): Water resources and their management in central Asia in the early twenty first century: status, challenges and future prospects. Environmental Earth Science 73 487-499 DOI 10.1007/s 12665-014-3789-1

Krätz D., Ibisch R., Avylush S, Enkhbayar G, Nergui S, Borchardt D. (2010): Impacts of Open Placer Gold Mining on Aquatic Communities in Rivers of the Khentii Mountains , North-East Mongolia. Mongolian Journal of Biological Sciences 2010 Vol.8 (1):41–50

Menzel L., Hofmann J., Ibisch R. (2011): Untersuchung von Wasser und Stoffflüssen als Grundlage für ein Integriertes Wasserressourcen Management im Kharaa-Einzugsgebiet (Mongolei). Hydrol. Wasserbewirtsch. 55(2):88-103

Minderlein S., Menzel L. (2015) Evapotranspiration and energy balance dynamics of a semi-arid mountainous steppe and shrubland site in Northern Mongolia. Environmental Earth Science 73:593-609 DOI 10.1007/s12665-0143335-1

Ministry of Environment and Green Development (MEGD) (2012). Integrated water management national assessment report volume II (2012) ISBN 978-99962-4-538-1

Nriagu J., Nam D.H., Ayanwola T.A., Dinh H., Erdenechimeg E., Ochir C., Bolormaa T.A. et al. 2012 High levels of uranium in groundwater of Ulaanbaatar, Mongolia. Science of the Total Environment DOI: 10.1016/j.scitotenv.2011.11.037

Pfeiffer M., Batbayar G., Hofmann J., Siegfried K., Karthe D., Hahn-Tomer S. (2015) Investigating arsenic (As) occurrence and sources in ground, surface, waste and drinking water in northern Mongolia. Environmental Earth Science DOI:10.1007/s12665-013-3029-0.

Punsalmaa B., Ydamsuren E., Dashdorj T., Tserendendev O., Batnasan D., Zandaryaa S. (2013): Water Quality of the Kharaa River Basin, Mongolia: Pollution threats and hotspots assessment, UNDP-GEF project "Integrated Natural Resource Management in the Baikal Basin Transboundary Ecosystem". Ulaanbaatar city.

Thorslund J., Jarsjö J., Chalov S.R., Belozerova E.V. (2012): Gold mining impact on riverine heavy metal transport in sparsely monitored region: the upper Lake Baikal Basin case. J Environ Monit 14(10):2780-2792. Doi:10.1039/c2em30643c

Heavy Metal Fluxes in the Rivers of the Selenga Basin

Galina L. Shinkareva, Nikolay S. Kasimov, Mikhail Yu. Lychagin

M.V.Lomonosov Moscow State University, Faculty of Geography

Abstract

In recent years, increasing attention is drawn to the Lake Baikal water quality due to the growth of human activities in its catchment area. A special attention is paid to the transboundary Selenga River as a major tributary of Baikal experiencing growing impact from industrial towns and mining activities. Our research aims to assess quantitatively the geochemical fluxes of heavy metals both in dissolved and suspended forms in aquatic systems of the Selenga River and its tributaries. The study is based on about 200 samples of river water and suspended matter collected during the high water period in August 2011 and the low water period in June 2012. The contents of heavy metals were obtained by ICP-MS and ICP-AES methods. The study revealed the significant increase of the heavy metal fluxes from the upstream (Mongolian) to downstream (Russian) part of the Selenga River basin. The geochemical fluxes experience considerable influence from Zaamar and Boroo mining areas, as well as Ulaanbaatar, Zakamensk and Ulan-Ude towns.

Keywords – geochemical fluxes, heavy metals, aquatic systems, Selenga River, Selenga delta, Lake Baikal, water quality, suspended matter

Introduction

The Selenga River is the biggest tributary of the Lake Baikal, which is the famous world's largest freshwater reservoir. It provides about 50% of the total river runoff into the lake. In recent years, the environmental state of the river is of a special concern due to the sharply increasing human activities in its basin. Selenga originates in the Hangai mountains of Mongolia and flows through the Republic of Buryatia (Russia) to the Lake Baikal. The Selenga River Delta accumulates more than a half of suspended matter that comes from the basin. The total catchment area of the Selenga River is 447 060 sq km and total length is about 1024 km, while length within Russian Federation is 409 km [Sorokovikova L.M. et al, 2000]. The main tributaries of the Selenga River in Mongolia are Tuul, Orkhon and Kharaa rivers.

There are numerous industrial and agricultural activities within the Selenga drainage basin, which influence the environmental state of aquatic systems. Main sources of anthropogenic activities in Mongolian part of the basin are the following: industrial towns Erdenet and Govi; metallurgy plant, carpet and woolen industries in Darkhan town; Ulaanbaatar city and gold mining areas in Boroo and Zaamar. Within the Russian part of the basin, there is also a number of pollution sources, such as Ulan-Ude town and W-Mo processing factory in Zakamensk town.

Our study aimed to determine the present pollutant levels, main features of their spatial distribution, and to assess quantitatively the geochemical fluxes of heavy metals both in dissolved and suspended forms in aquatic systems of the Selenga River and its tributaries.

Materials and methods

The study is based on the field data collected in the Selenga River basin during summer periods of 2011-2012 in both Mongolia and Russia (fig. 1, table 1). In 2011, the fieldwork was held in August to estimate migration of heavy metals during summer rain floods. In 2012, the sampling was done in June during the low water period.

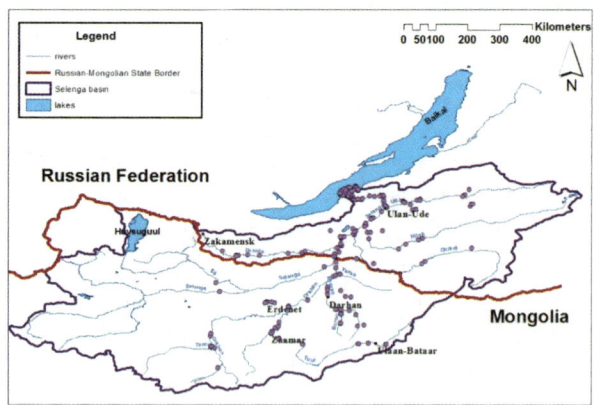

Figure 1. Study area. Sampling points

Subject	Number of samples			
	Russian part (except delta)	Russian part (delta)	Mongolian part	Total
River water	87	43	65	195
Suspended matter	88	45	60	193

Table 1. Field sampling within the Selenga River basin in 2011-2012

At each point some water characteristics were measured in situ from the boat with HANNA field equipment, e.g. pH, Total Dissolved Salts (TDS), Redox potential (ORP). Water was sampled in 2l and 5l bottles. The water samples were filtered with Millipore Vacuum Pump (230v 50Hz) through 0.45 mic filters. The filtered samples were preserved with 1 ml of strong HNO_3 and stored in 15 ml tubes with a screw cap. The filters with suspended matter were dried at room temperature.

Later on dried filters and preserved water samples were taken to Moscow Federal state unitary enterprise «All-Russian scientific-research institute of mineral resources named after N.M. Fedorovsky» where they were analyzed with ICP-AES and ICP-MS methods. As a result, the concentrations of 71 elements were obtained. Among all of them 13 main contaminants were defined and analyzed with different toolkits: ArcGIS 10, STATISTICA 8, Microsoft Office 2007.

On certain sites some hydrological characteristics, such as velocity, turbidity and morphometric parameters of channel were determined also. The water velocity was measured with hydrometric current meter ISP-1 at the one-fifth depths of each width increment, turbidity with HACH 2100P Turbidity Meter. The total water discharge was calculated by multiplying the discharge flow velocities with cross-sectional areas of the rivers. The total sediment discharges were calculated by multiplying sediment-load velocities with average suspended sediment concentration values and cross-sectional areas of the rivers.

For all the considering elements fluxes of heavy metals per day in dissolved (W_c) and suspended (W) forms were calculated:

1) For dissolved forms: $W_{ci} = Q \cdot C_i \cdot 86400 / 10^9$, where W_{ci} – flux of i-element in dissolved form (kg per day), Q – water discharge (cbm per second), C_i – concentration of i-element in river waters (mkg per cbm), 86400 – conversion factor seconds per day, 10^9 – conversion factor mkg per kg.

2) For suspended forms: $W_i = R \cdot C_i \cdot 86400 / 10^6$, where W_i – flux of i-element in suspended form (kg per day), R – discharge of suspended matter (kg per sec), C_i - concentration of i-element in suspended matter (mg per kg), 86400 – conversion factor seconds per day, 10^6 – conversion factor mkg per kg.

Results and Discussion

All the results were compared with various average values. For the content of heavy metals (HM) in river water, the maximum permissible concentrations (MPC) for fishery water subjects were used. Values of HM in suspended matter were collated with the average composition in upper continental crust (UCC) according to K. Wedepohl [Wedepohl H.K., 1995] and average composition of suspended matter in river water [Savenko, 2006]. In the course of data interpretation, the greatest interest was aroused by such HM as V, Cr, Mn, Fe, Co, Ni, Cu, Zn, As, Mo, Cd, Pb, and U, which are mostly discussed in the article.

River Water

Heavy metal concentrations in the water samples vary from the season to season (table 2), but some similarities can be defined. For example, average concentration of HM dissolved forms during the low water period of 2012 decreased in comparison with the period of floods 2011.

Element	MPC[1]	2011[2]	2012[2]
V	1	1.93	2.69
Cr	70	8.16	13.41
Mn	10	28.32	59.95
Fe	100	433.22	985.86
Co	10	0.47	0.57
Ni	10	1.42	2.00
Cu	1	6.67	2.74
Zn	10	35.00	10.05

As	50	3.12	8.69
Mo	1	21.20	2.57
Cd	5	0.29	0.04
Pb	6	7.28	1.51
U	-	2.50	3.42

[1] - according to [An Order of Russian Federal Agency for Fisheries];
[2] – our data.

Table 2. Average concentration of HM, mkg/L, in field seasons 2011 and 2012, and Maximum Permissible Concentration for fishery water subjects (MPC).

Comparing HM concentrations with the maximum permissible levels we can see that there are elements, average concentration of which is higher than MPC in both cases: V, Mn, Fe, Cu, Zn, Mo.

Geochemical fluxes of dissolved HM generally increase towards the delta area. The Kharaa River upstream the discharge of the Boroo River had the lowest fluxes of both dissolved and suspended HM. In the main stream of the Selenga River near Russian-Mongolian state border right after the confluence of Selenga and Orkhon rivers and at the apical part of the Selenga delta the highest fluxes of metals in both forms were found in 2011 (fig. 2).

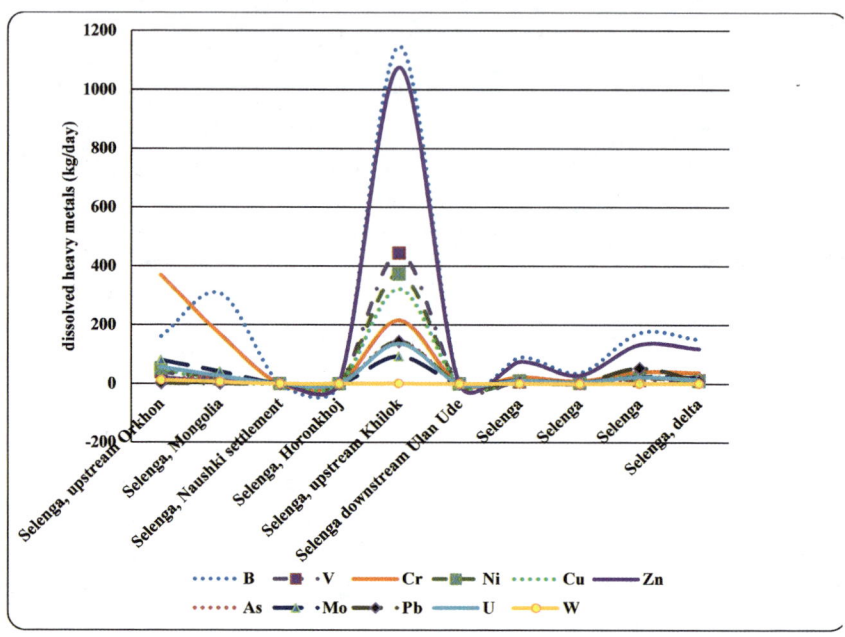

Figure 2. Fluxes of dissolved heavy metals along the Selenga River (2011), kg/day

At the Figure 2 we can see also the influence of Dzhida and Chikoj rivers: up-stream the Khilok river and at the same time downstream the Dzhida and Chikoj rivers there is a huge increase of fluxes of all elements: from 200 times for Zn (5,4 ÷ 1075,7 kg per day) to 7500 times for V (0,06 ÷ 444,61 kg per day). There are two major reasons for such a big difference: firstly the Dzhida river provides big inflow of heavy metals from the Zakamensk town and its mining and processing plant. The second reason is that the water discharge increased at this point by 1,2 times (from 679,4 cbm at the Russian-Mongolian State bor-der to 830,0 cbm downstream Dzhida and Chikoj rivers). Also the groundwater discharge can impact a lot.

During the low water period of June 2012 the largest fluxes of dissolved heavy metals were found in the Selenga River downstream Ulan-Ude town (fig. 3) and at the beginning of the delta.

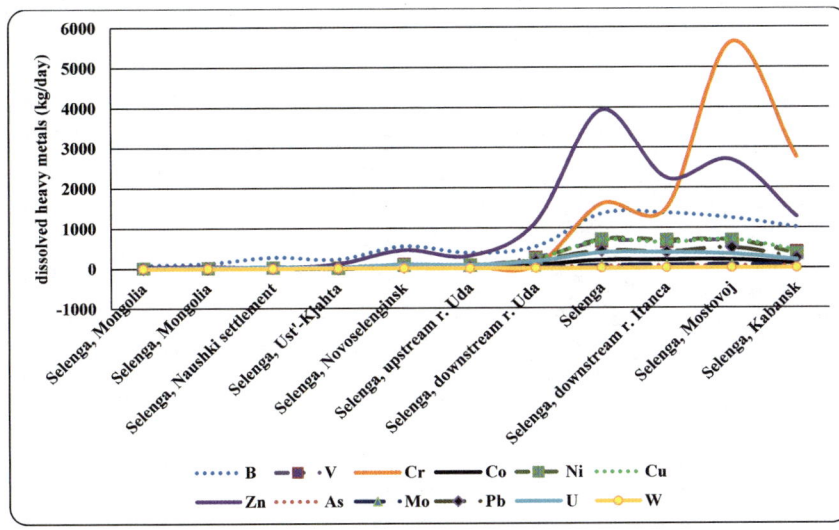

Figure 3. Fluxes of dissolved heavy metals along the Selenga River (2012), kg/day

To estimate the geochemical impact of Ulan-Ude town we have compared the dissolved HM fluxes in the Selenga upstream and downstream the confluence of the Uda River. The fluxes increased significantly: by 1,2 times for As, Cr; 1,4 for V; 2,1 for U; 3 for Mn and Ni; 3,7 and 3,8 for Zn and V respectively; 4,2 and 4,3 for Co and Fe; 5,8 for Pb; 10,2 for Cu. Only for dissolved Mo the flux slightly decreased (by 1,2 times).

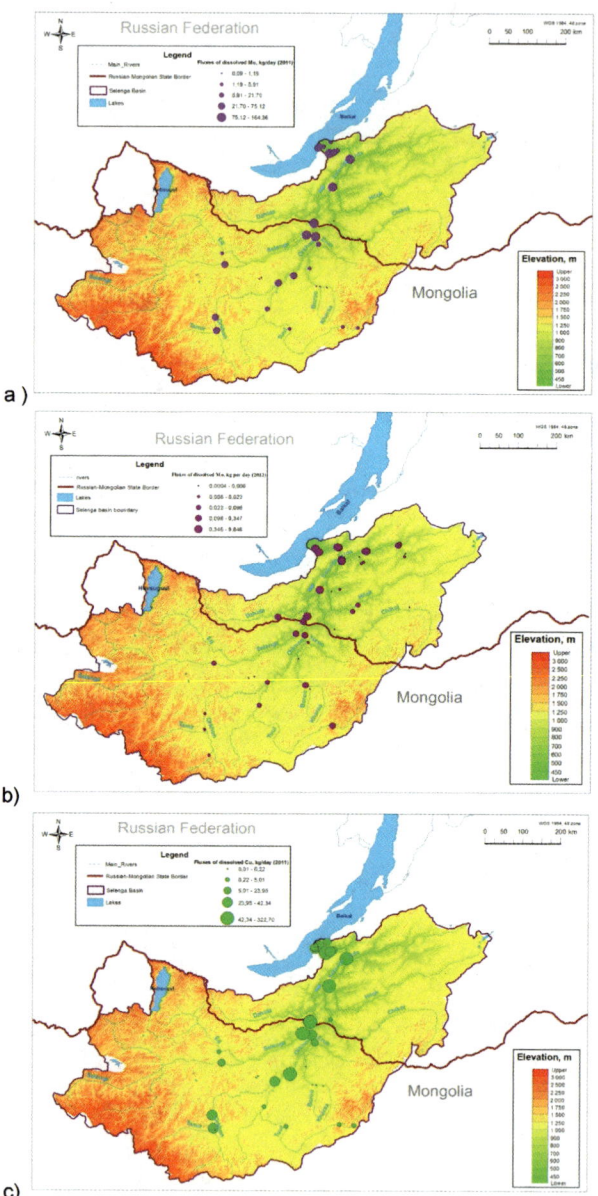

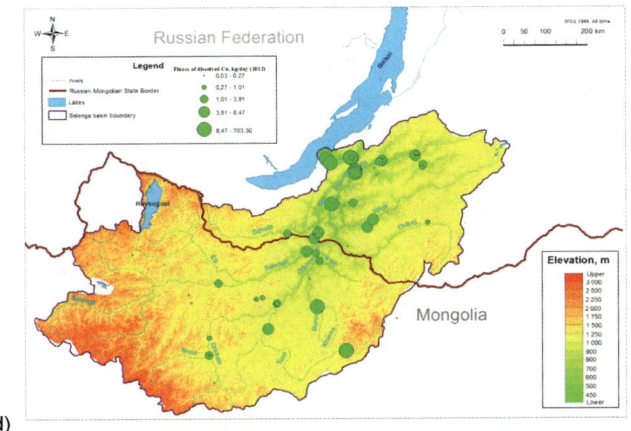

d)

Figure 4. Fluxes of dissolved Mo in 2011 (a) and 2012 (b) and dissolved Cu in 2011 (c) and 2012 (d) in kg/day

Some peculiarities of HM fluxes within the Selenga basin are shown on the Fig.4. During the low water period 2012 the main flux of Cu in the Mongolian part of the basin went from the upstream Tuul river (17,9 kg per day) while the Orkhon river didn't contribute that much (less than 1 kg per day). The copper flux changed significantly downstream the Selenga River itself. In the upper Mongolian part it was rather small (4,6 kg of dissolved Cu per day), than it increased up to 8,3 kg per day after the confluence with the Orkhon River. In the Russian part of the river the Cu flow arouse up to 32,9 kg per day after the confluence with Dzhida river. However the largest contribution into the flux of dissolved Cu was made by Ulan-Ude: downstream the town it increased from 22,8 to 233,2 kg per day. Towards the Selenga River delta the flow increased up to 703,3 kg per day.

The similar pattern of the spatial changes during the low water period was found for Mo fluxes. However during the high water period (2011) fluxes of dissolved Cu and Mo in Orkhon river and Selenga itself were close (40-50 kg per day for Cu and 60 kg per day for Mo). These values redoubled downstream Russian-Mongolian State Border (126,2 and 158,5 kg/day respectively) slightly changed towards the Selenga delta (134,7 and 142,2 kg/day).

Suspended Matter

Heavy metal content in the suspended matter (Table 3) were compared with the average composition of the upper continental crust [Wedepohl, 1995] and average composition of suspended matter in World's rivers [Savenko, 2006]. After the comparison, we can say that concentrations of Mn, Ni, As, U are above both estimations for all seasons.

Element	UCC[1]	World's average[2]	2011[3]	2012[3]
V	53	120	151	86
Cr	35	85	69	48
Mn	527	1150	2308	1871
Fe	30890	50300	51382	34945
Co	11,6	19	29	18
Ni	18,6	50	50	28
Cu	14,3	45	206	35
Zn	52	130	226	80
As	2	14	25	25
Mo	1,4	1,8	5,3	1,7
Cd	0,102	0,5	1,42	0,24
Pb	17	25	209	0,9
U	2,5	2,4	8,4	5,7
[1] - according to [Wedepohl, 1995]; [2] - according to[Savenko, 2006]; [3] - according to field measurements.				

Table 3. Average concentration of HM, mg/kg, and average values for the suspended matter

In August 2011, the major flow of the suspended matter in the Mongolian part of the basin come from the Orkhon River (fig. 5). For instance, the flux of suspended Cu in the Orkhon River upstream confluence with the Tuul River was found as much as 500 kg/day while downstream it changed to 188 kg/day. Tuul River in a less extent is responsible for the pollutant flows. However downstream the Zaamar mining area, which is located on the Tuul river, fluxes of suspended metals increase as well, e.g. for Ni by 2,7 times, Sn by 3,8 times, Mo, Cd and Sb by 1,5 times. In June 2012, the largest suspended matter flow was found at the Selenga River downstream the confluence with the Itanca River, and further downstream to the beginning of the Selenga delta, where for Cu it was equal to 1344 kg/day and for Mo 264 kg/day. The Ulan-Ude also contributed a lot: downstream this town flows of suspended Cu and Mo in 2012 low water period changed from 25,6 and 8,4 kg/day to 209,4 and 55,8 kg per day respectively.

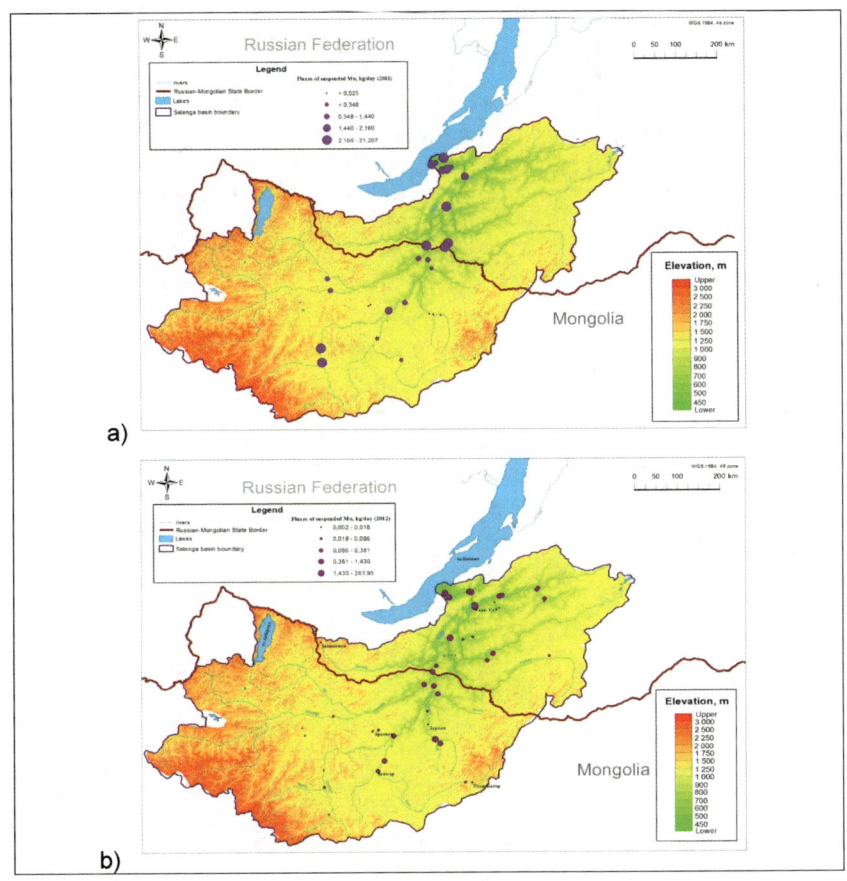

a)

b)

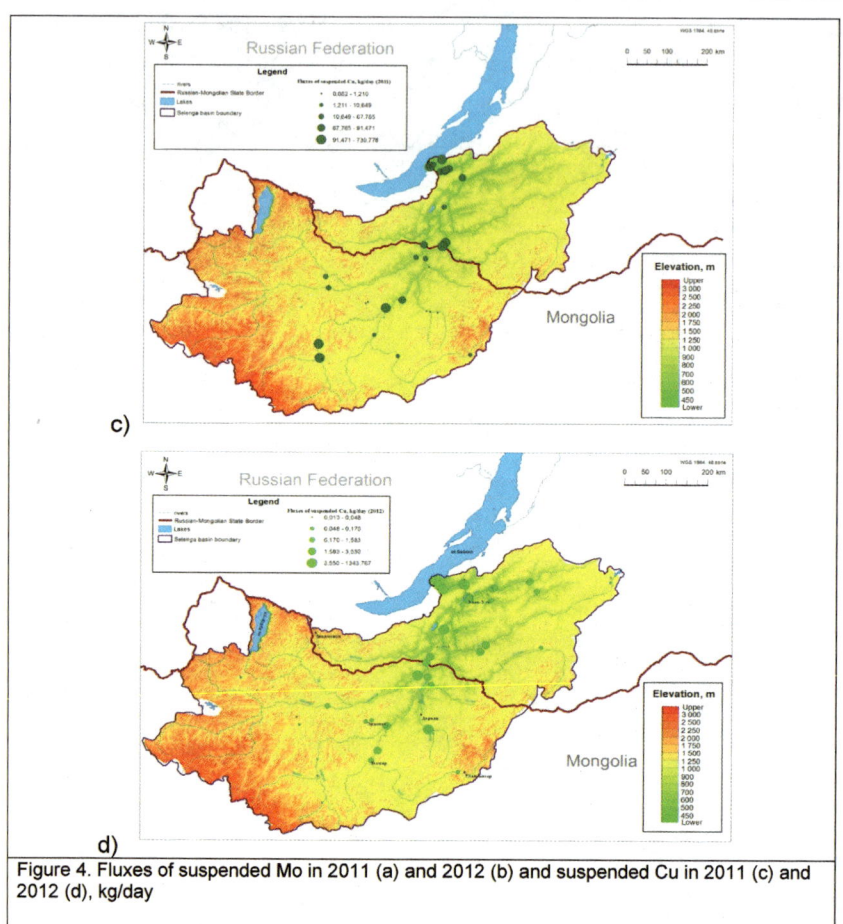

Figure 4. Fluxes of suspended Mo in 2011 (a) and 2012 (b) and suspended Cu in 2011 (c) and 2012 (d), kg/day

The seasonal changes of HM fluxes were found significantly different for the Selenga River tributaries. For example, the Tuul River during the low water period 2012 was characterized with higher flows of dissolved Mn (in 200 times), Zn, Fe (50-40), Cd, Cr, Co, Cu, Ni (10-3), and suspended V and Mo (3-2 times). The HM fluxes in the Orkhon River were found larger in the high water period: dissolved Mo and Pb (more than 1000 times), U (100), Cu, As, Ni, V (50-10), Fe, Co, Zn (10-2 times). For the suspended forms the difference is even higher: U (7600 times), Pb, Cd, As, Cu, Ni, Co, Cr, Zn (300-100), Mn, V, Fe, (100-50).

On the contrary, in the Khan Gol River the HM suspended fluxes changed slightly: Mo in 11, Zn in 6 times. For the dissolved formes the difference is higher: Fe, Mn (1000), Zn, Cu, Co (40-10), Cd, Ni, V (5-3). For the Kharaa River the difference in suspended flows between the two seasons was distinguished only for U and V (in August 2011 the flow was higher more than in 10 times), and also Cr, Cd, Pb (2-3); the HM dissolved form fluxes were found larger during the low water period (June 2012) for Mn, Fe (more than 100 times), Co, Zn, Cd, Ni, Cu (10-2).

Conclusions

Analysis of the heavy metal levels in rivers of the Selenga basin showed that average content in water of V, Mn, Fe, Cu, Zn, Mo is above MPC fishery values. Content of Mn, Ni, As and U in the suspended matter is higher than average composition of upper continental crust and average composition of suspended matter in World's rivers.

Along the main stream of the Selenga River the fluxes of both dissolved and suspended forms of heavy metals decreased in low water period in the Mongolian part of the basin (except for dissolved Mn, Fe, Zn, and Cd). In Russian part of the basin fluxes of dissolved and suspended V, Cr, Mn, Fe, Co, Ni, Cu, as well as dissolved Cd and suspended Zn, As, Mo increased in the low water period from 2,5 to 50 times.

Geochemical fluxes of dissolved HM increased towards the Selenga River Delta from 2 (V, Ni, Mo) to 485 (Mn) times during high water period and from 1,4 (Cd) to 378 (Cr) times for low water 2012. Flows of suspended forms had grown form 1,7 - 2,0 (Mn, Cr, Co, Ni, Cu, U) to 25,7 (Cd) times in 2011 and from 32 (V, Cd, Pb) to 358 (Mo) times in 2012.

During the period of summer rain floods Orkhon River provided the main runoff of HM in suspended and dissolved forms from Mongolian side and around Russian-Mongolian State Border it increased its maximum. Chikoj and Kiran rivers contributed to the dilution of Selenga River but towards the Selenga delta area this flow started to rise again. During the low water period HM fluxes in both forms grew along the stream of the Selenga River and reached the maximum values right before the apical part of the Selenga delta.

References

Sorokovikova L.M., Sinjukovich V.N., Golobokova L.P., Chubarov M.P. "Formation of ion runoff of the Selenga river in the modern conditions." Water resources and conditions of water facilities. – 2000. - № 5. – Vol. 27. – p. 560-565. (in Russian)

Hans Wedepohl, K. "The composition of the continental crust." Geochimica et cosmochimica Acta – 1995 – № 7. – Vol. 59 – pp. 1217-1232.

Savenko V.S. "Chemical composition of suspended matter in World's rivers". Moscow. GEOS – 2006. – 174 p. (in Russian)

An Order of Russian Federal Agency for Fisheries from 18.01.2010 № 20 "On approval of the water quality standards for fishery water bodies, including the maximum permissible concentrations of harmful substances in the waters of fishery water bodies". Federal Agency for Fisheries. (in Russian) – URL: http://fish.gov.ru (10.05.2012).

Linking Catchments to Rivers: Flood-driven Sediment and Contaminant Loads in the Selenga River

Chalov Sergey R.[1] and Romanchenko Anna O.[1]

[1] Lomonosov Moscow State University, Moscow, Leninskie gory, GSP-1, 119991.

Abstract

The present study is particularly devoted to link geochemical flux and climate and land use changes in the Selenga river catchment. Results indicate that high sediment loads were reported both for altered and natural rivers. Reported multi-decadal declines in sediment loads in the downstream part of Selenga River can be attributed to the abandonment of cultivated lands and changing hydroclimatic factors, such as in particular climate-driven decrease of water flows and intensified water use for irrigation purposes. Empirical sediment rating curves show that a series of peak flow events during spring and summer contribute to the main part (up to 98 %) of the annual sediment and pollution loads. The highest contribution of flood sediment load was obtained for the particular wet years, the lowest - for the dry, which is generally reflects the increase of water runoff during high floods in annual flow. Whereas sediment flows were connected with the hydroclimatic conditions in the catchment, elemental composition of the mass flows mostly relate to the soil/petrologic conditions. With the exception of small impacted rivers where water quality impacts associated with mining were found, the formation of elemental compositions and sediment-associated chemical constituents generally reflects catchment characteristics

1 Introduction

Recent studies point to a strong coupling of river discharge, climate oscillations and land use (Syvitski, 2003). Human-induced changes have significantly altered both the composition and magnitude of allocthonous material entering rivers. The latter leads to the significant shift in geochemical fluxes and water quality. As far as sediment transport rates are also a function of sediment avail-ability (Asselman, 1999), the climate oscillations and land use changes have a profound impact during flood events when physical and biological processes operate to increase matter export. In this perspective close interdependence among climate, land use, vegetation cover density, and erosion rates remains the key questions in both global and regional estimates of the sediment and contaminants loads (fig. 1).

Fig. 1. Schematic representation of the controlling factors of catchment/river interface (red lines in-dicate sediment delivery during floods events, green dotted lines – during low-water periods)

Changes in sediment availability during the floods are resulted in so-called hys-teresis effects. Clockwise (positive) hysteresis loops (type II), anti-clockwise hysteresis (type III) as far as single (direct, no hysteresis effect) (type I) or com-plicated type (IV) of S=f(Q) relations can occur (Gellis, 2013). The export of land-derived constituents from drainage basins as a result of episodic dis-charge could exerts first-order control on the cycling of sediment and contami-nants loads in rivers (e.g. Zwolsman, 1997; McKee et al. 2004).Typically (table 1) studies on flood impacts on in the context of finding the bridge between catchment and rivers are based on:

- sediment rating curves and hysteresis effect analyses
- intensive detailed basinwide investigations of the small catchments
- detailed investigations on the long-term observational data for the outlet stations in the large catchments as a surrogate measure of catchment to river interface

With some exceptions (e.g. Asselman, 1998) large-size river basins are chara-cterized by the data from outlet stations, whereas chemical elements behavior during hydrological event is much less studied (e.g. Audry et al., 2004, Horo-witz et al, 2008, Ollivier et al., 2011). Conversely, much more detailed studies in terms of the observational data have been done for small catchments (e.g. Roussiez et al, 2013; Gellis, 2013) (table. 1). For both scales relatively small amount of studies have addressed the composition and magnitude of the material supplied during short-lived events.

Study area and scale	Methods	Datasets (station, time, variables)	Catchment properties	Reference
10 sub-catchments of the Geba river, northern Ethiopia (F=from 121 to 4592 km²)	Basinwide sediment rating curves analyses, determination of grain size distribution (sieve-pipette method)	41 monitoring stations, 2004 – 2007 during rainy season (July-September), pressure measurement every 10 min (TD-diver)+manual measurements of flow depth, runoff discharge and sampling of suspended sediments (2-3 times per week in 2004, in 2005-2007 – daily)	Topography, size, land use, vegetation cover, lithology	Vanmaercke, 2010
Rhine catchment between Kaub and German-Dutch boarder (165000 km²)	Basinwide sediment rating curves analyses, "supply-based" model (computes sediments transport as a function of water discharge and the amount of sediment in storage)	Stations along Rhine river and its main tributaries, 1975-1990 - daily discharge and suspended sediment concentrations	Base of slopes, drainage area, distance from river source, travel time of the sediments	Asselman, 1998
4 catchments of Puerto Rico (F=from 3,5 to 20 km²)	Sediment rating curves analyses using Factor analysis and stepwise regression on the factor scores.	1 site per catchment characterized by various land use storm-generated sediment loads and concentrations	Type of land use	Gellis, 2013
Wadi Sebdou catchment (256 km²)	Graphical analysis method based on features of hysteresis loops	31-year period (1973-2004) at the outlet of basin	Seasonality of flood types and sediment sources	Megnounif et al, 2013
16 in Northern Siberia and Far East of Russia (large and medium rivers)	Sediment rating curves analyses, plots grouped over the phases of water regime	27 gages (1-2 station per catchment) with chronological plots of S=f(Q) relationship	Duration of events (spring flood, freshet period), sediment sources, seasonal permafrost, topography, soil types, travel time of the sediments	(Tananaev, 2011)
Selenga catchment (F = 447000 km²)	Basinwide accounting segment	150 grab sampling sites around the catchment with daily water discharges and more than 50 chemical variables; 7 monitoring stations	Topology, land Use and land cover, population density, climatic variables	Present study

Table 1. Some examples of the regional studies of land-derived constituents during floods

The key task still remains open is to disentangle the influence of climate change and related changes in floods and episodic discharges from that of other changes in catchment condition (firstly land use drivers). Less is known about flood sediment transport magnitude and hysteresis effect connections with magnitude and structure of geochemical fluxes.

The present study is focused on detailed analyses of geochemical fluxes and the role of single floods in suspended sediment dynamics due to climatic, hydrological and land use impacts within relatively large Selenga river basin (table 1). The Selenga River, which originates in Mongolia, contributes about 50 % of the total inflow into Lake Baikal. The catchment is still lacks of any type of soil and water conservation and sediment control programmes as far as reservoir constructions which are in the global perspective cause decreased sediment fluxes (e.g. Walling, Fang, 2003). The main anthopogenic inputs are related to mining, industrial and agricultural activities within the Selenga drainage basin. At the same time, the region is reported to experience the warming trends with acceleration since the 1970s (Unger-Shayesteh et al., 2013), what has a profound impact on the components of hydrological system in the area. Elevated sediment-associated chemical concentrations were reported for the area (Chalov et al, 2014).

In particular, this study aims at (i) long-term changes of hydrocilmatic drivers of sediment delivery into river channel, (ii) investigating the impacts of hydrological peak-flow events on total annual load contributions and their implication on geochemical fluxes, (iii) determining importance of export of land-derived constituents and related human activities on the cycling of sediment and contaminants loads in rivers

2 METHODS

The study is based on data from the national gauging network of the Selenga River Basin, which is implemented by the Russian and the Mongolian hydrometeorological surveys for their corresponding parts of the basin. The long-term hydrological changes in the selected rivers were calculated based on reference period 1975-1995 for the average values of 1996-2011:

$$\Delta Q = \frac{Q_{1996-2011}}{Q_{1975-1995}}$$

Regional climatic modelling of land use impact on air temperature and precipitation was done based on COSMO-CLM (Consortium for Small-scale Modeling) tool (Böhm et al, 2006) with a spatial resolution 14 km. The modelled approach was based on the comparative study of predicted precipitation and temperature in in Central part of Selenga catchment (Northern Mongolia) simulated for 2 scenarios:

1. Scenario of modern landscapes to predict actual recent values of precipitation P_0 and temperature fields T_0
2. Scenario of landscapes extremely disturbed by mining operations to predict precipitation $P_{disturbed}$ and temperature $T_{disturbed}$ fields in altered conditions. The areas of disturbed lands were determined according to areas available to mining approved recently by Mongolian Government (fig. 2).

The difference of modelled values between undisturbed and disturbed conditions were calculated as

$$\Delta P = \frac{P_0}{P_{disturbed}} \quad \text{and} \quad \Delta T = \frac{T_0}{T_{disturbed}}$$

Suspended sediment monitoring data included daily averages for the period 1970-2010 which were used to obtain S=f(Q) relationships (table 2) between daily suspended sediment concentrations S and daily water discharges Q for the flood season which covers both melting period (April-May) and rainfall floods (occurred in June-August). Each flood event was characterized by the particular relationships. The representative equation for the given gauging station was used in further analyses to estimate contribution of flood season to the annual sediment yield.

River-station	Type of S(Q) relation	Relationship	
		rising limb	falling limb
Selenga - Mostovoi	IIIb	$S = 0.043Q^{1.05}$	$S = 0.0002Q^{1.7}$
		$R^2 = 0.84$	$R^2 = 0.64$

Chikoy -Gre-myachka	IIIb	$S = 0.011Q^{1.58}$	$S = 0.0027Q^{1.7}$
		$R^2 = 0.78$	$R^2 = 0.63$
Khilok - Khaylastuy	IIa	$S = 6.81e^{0.009Q}$	
		$R^2 = 0.68$	
Uda – Ulan-Ude	IIIb	$S = 2.29Q^{0.79}$	$S = 0.02Q^{1.66}$
		$R^2 = 0.71$	$R^2 = 0.79$
Khara - Burunkhara	Ib	$S=17.6Q+87.3$	
		$R^2 = 0.73$	

Table 2. Representative S=f(Q) relationships during flood season (April to August)

For the detailed investigation of geochemical patterns novel screening campaigns were conducted in June–September 2011-2014 in both the Russian and the Mongolian parts of the Selenga river basin. The special focus of the study is on the episodic rainfall discharges and associated geochemical fluxes which were analysed on the data from 3 monitoring stations located in the Upper Orkon river (near Kharokhorin), Tuul river (near Ulan-Baatar) and Khara river (near Burunlhara) (fig. 1). Water discharge and water sampling were performed during 3-4 weeks campaigns in 2011, 2012, 2013 and 2014. To record suspended sediment data, 2000-4000 mL plastic sampling bottles were used to collect water samples manually both with optical turbidity measurements by Hach 2100P during the rising and falling limb of a storm hydrograph. The water samples were filtered through pre-weighed filter papers of pore size 0,45 μm. Filter papers were oven dried, weighed, and suspended sediment concentrations calculated gravimetrically. Moreover, the water samples were also collected during rainfall–runoff events in 2012 at intervals of 5 minutes at Khara river (near Burunkhara) by YSI automatic sampler operated by Momo project (Karthe et al, 2014).

Discharge and suspended sediment concentration (SSC) data were combined to yield estimates of daily and monthly water discharges, suspended load averages at more than 150 locations. All samples (suspended and streambed sediments and filtered water) were analyzed for 62 elements by inductively-coupled plasma mass spectrometry ICP-MS (ICP-AES) using a semi-quantitative mode and a 10-fold automated dilution during the analysis. Elemental analyses were conducted on the filtered samples without additional treatment. For a fully quantitative analysis, the instrument is calibrated with a series of known standards for each element. Corrections are applied for potential interferences, and more comprehensive quality assurance/control measures are performed for each element.

The spatial variability of geochemical fluxes was evaluated based on environmental surveys conducted in 2011- 2014. The surveys targeted sites located along the Tuul River (T), the Orkhon River (O), the Eg River (EG), the Yeroo River (ER), the Khangal River (H), the Selenga River (S) and the Kharaa River (Hr) in Mongolia (Fig. 1). In Russia the observational sites were located along the main stem of the Selenga River (S) and its main tributaries – Dzhida (D), Temnik (TM), Chikoy (CHK), Hilok (HK), Orongoy (OR), Uda (U), Itantsa (IT), Kiran (KR), Kudara, Zheltura (G), Udunga (UD), Suhara (SH), Tugnui (TG), Menza (MZ), Buy, Bryanka (BK), Ilka (IK), Chelutay, Kurba (KB), Kodun (KD), Kizhinga (KG), Ona (Fig. 2).

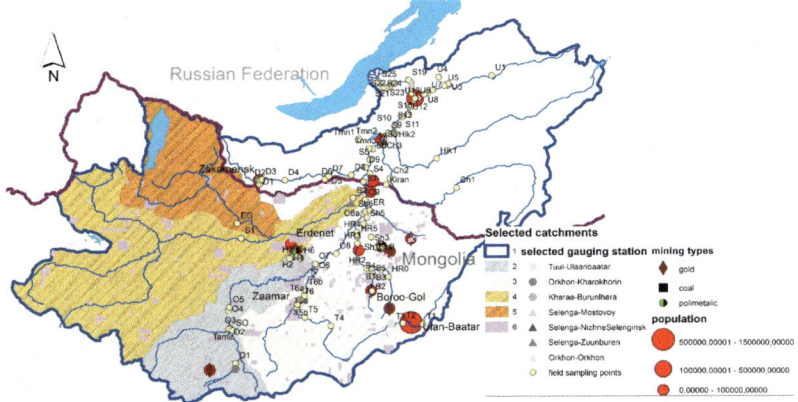

Figure 2. Selenga River Basin showing sampling sites, major mining sites and main subcatchments (1 – whole Selenga catchment. 2 – Orkhon river, upper part; 3 – Tuul river; 4,5 – upper Selenga River with Eg River and Khovsgol Lake); 6 – areas available to mining in Mongolia (unpublished data from E. Simonov)

3 DISCUSSION AND RESULTS

3.1 Hydrocilmatic drivers of sediment delivery

Climate variability and change is one of the major driver of hydrological trends in dry lands. In Mongolia, air temperature has increased by 1.8⁰C since the 1940s, and precipitation decreased in some parts of the country, including the Western slopes of the Khentii mountains. The latter is regarded to be the main reason of long-lasting low water period (since around 1989) that is reported for the Selenga River. The runoff records for downstream Selenga (Mostovoy) shows statistically significant downward trend from 903 m³/s (for the period since 1941 to 1982) to 888 m³/s (for the period since 1983 to 2011), or even more drastic for the recent decades – from 940 m³/s (for the period since 1975 to 1995) to 689 m³/s (for the period since 1995 to 2011). The mean annual discharge at some gauging stations within the Mongolian part of the basin) demonstrated even more significant decrease during the last decades, from an average of 53,3 m³/s during the period 1975-1995 to only 18,1 for the years 1996-2011 at the upper Orkhon river (Orkhon-Orkhon) and from 35,3 m³/s to 14,9 m³/s for the Tuul river at UlaanBaatar respectively. The changes were caused mostly by intensive rainfall floods which determined high water period in the early 1990 and late 1970-1980. This decrease primarily caused a reduction in precipitation and an increase in evapotranspiration during that period, though intensified water use for irrigation purposes may have contributed as well.

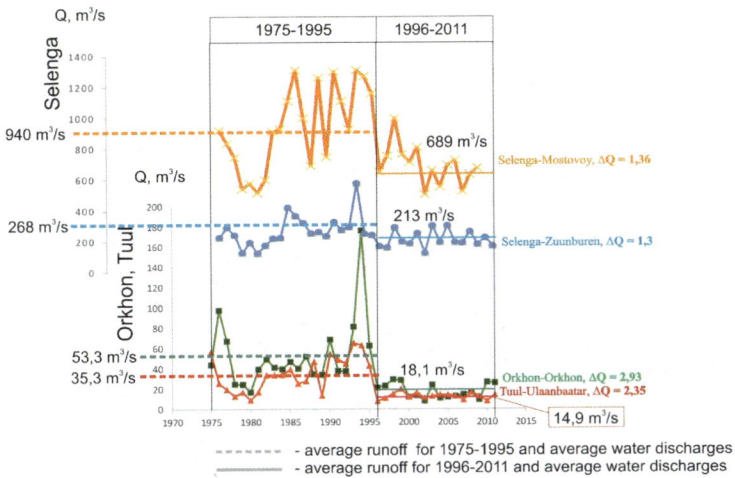

- - - - - - - - average runoff for 1975-1995 and average water discharges
———————— - average runoff for 1996-2011 and average water discharges

Figure 3. Long-term changes of average runoff in Selenga catchment (location of the gauging station – see fig. 2)

Decrease of observed mean annual Q since 1990 is largely due to decreased peak discharge during summer. Mentioned climatic variability have been significantly changed the maximal discharges over last decades (fig. 4) and the flood contribution became smaller than in the middle of XX century (average annual difference between maximal and minimal discharges Qmax – Qmin decreased from 3740 to 2920 m³/s). This correspond to the highest warming rate in the Selenga River Basin during the latest 20-year period (1989–2009; 0.048 °C/year). On the other hand, land disturbance, especially by mining operations, could significantly influence water balance components due to changes in regional evaporation patterns. The results of COSMO-CLM model application indicated spatial redistribution of both precipitation and temperature field with maximal shifts in annual summer precipitation ΔP=300 mm. The most significant shifts will occur during dry years.

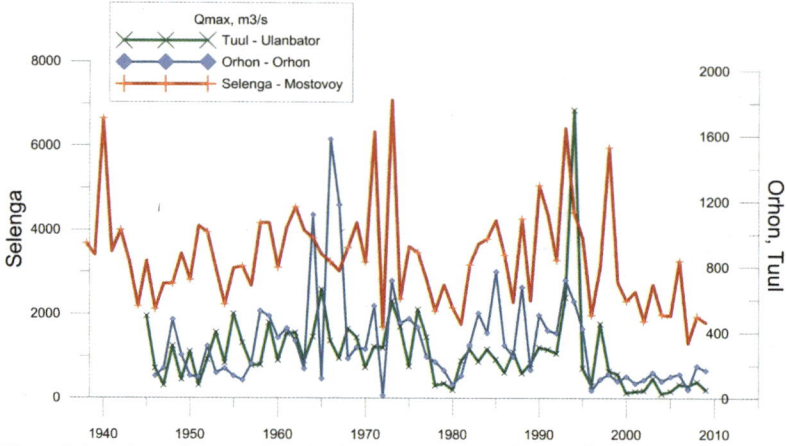

Figure 4. Long-term changes of maximal annual discharges Qmax (m³/s) at Tuul, Orkhon and Selenga river

Following mentioned changes in climatic conditions, there has been a substantial decline in sediment yield of Selenga River (from 5832 t/day to 3015 t/day) and its main tributaries in the Russian part of the river basin. In the upper part of the basin where an absence of routine monitoring of sediment loads precludes statistical analyses of the sediment trends, the few evidences of the sediment yield decrease have been seen based on the comparison between SPM concentrations measured during the campaigns of 2011-2014 and historical field campaigns of 1934-1936 (Kuznetsov, 1955). At the upper part Selenga above the Eg River (S-1) SSC varied between 1.2 mg/l (20th February) to 1193 mg/l (8th August) in 1934, whereas in 2001 it was 11.5 mg/l (18th-24th August). During our field campaigns SSC varied from 9.51 mg/l (16th June, 2012) to 114 mg/l (2nd August, 2011). At the confluence of Orkhon and Tuul, a 65-fold increase of SSC was observed during summer floods for the Tuul River (T-9) (from 11 mg/l at the 19th October, 1934 to 716 mg/l at the 26th August, 1934) and 43-fold for the Orkhon River (O-6) (from 23.3 mg/l at the 17th June, 2012 to 1000 at the 7th May, 1934).

Taking into account the CMIP 5 ensemble mean projections expected continued T increase by as much as 5°C until the end of the XXI century (Törnqvist et al, 2014), increased runoff in near in the future is indicated. This implies shifts in sediment transport patterns particularly during event conditions. Among the main driving forces of the water and sediment transport are shifts in soil temperature and moisture which exert a strong control on soil aggregate stability, and thus on soil erosion intensity; permafrost thaw which likely to continue. Recent mining impacts on sediment loads were seen mostly during relatively short hydrological events, during which an intensified slope wash near floodplain mining activities could flush large amounts of turbid water into the river. A significant suspended sediment load increase was reported for the Tuul River at sites located near the Zaamar placer gold mining area. During the flood season in 2011, a 2.3-fold sediment load increase from 307 to 710 t/day was recorded. During base flow periods in 2012, a 1.2-fold increase from 115 to 143 t/day was detected.

3.2 Impacts of hydrological peak-flow events

Long-lasting flood season typically has a significant effect on the annual sediment loads. Up to 98 % of annual sediment loads in the river are transported during the flood season (April to August) (table 3). The highest contribution of flood sediment load was obtained for the particular wet years, the lowest - for the dry, which is generally reflects the increase of water runoff during high floods in annual flow. The temporal variability of sediment discharge within individual basins indicates that maximal role of flood given sediment transport events relates to small rivers. Changes in sediment transport rates correspond with changes in SPM size which are probably related to the intensive slope wash during rainfall events that dominate the flood period. They may furthermore be related to seasonal changes of vegetation coverage and properties, temperatures (that influence the concentration of suspended matter), colloids, and organisms.

River-station	Year	Flood contribution		Duration of period, days/% of year
		R, tons	R/W % of year,	
Selenga - Mostovoi	1976(dry)	2720000	98	183/50%
	1967(average)	3120000	82	183/50%
	1973(wet)	13900000	99	264/67%
Chikoy -Gremyachka	1972(dry)	28100	52	183/50%
	1963(average)	138000	89	183/50%
	1962(wet)	431000	99	183/50%
Khilok - Khaylastuy	1977 (dry)	43300	72	188/52%
	1967(average)	87900	94	266/73%
	1973(wet)	345000	99	217/59%
Uda – Ulan-Ude	1977(dry)	30300	96	183/50%
	1963(average)	104000	97	216/59%
	1973(wet)	267000	99	218/60%
Khara - Burunkhara	2002(dry)	4340	97	189/52%
	1992(average)	27900	97	245/67%
	1990(wet)	38300	97	253/69%

Table 3. Flood contribution (R, tons and % of the year) to annual sediment yield W for the selected gauging station in Selenga river catchment

Variations of the sediment load during short-lived hydrological events (rain floods) can have even more significant effect on sediment transport patterns and depend on different synoptic situation and geological and geomorphological conditions in the catchments. During June-August 2012 rain floods in the downstream of the Khara river (KH-4) were relatively small (maximal change in the sediment concentrations in 2012 within two adjacent days was 22%). In the upper mountain part of the catchment (KH-2) the heavy rains when the total precipitation between the 20th and the 22nd of June was 28.1 mm (7 % of the annual value) lead to the SSC increase from 13.3 to 518 mg/l. The same effects of rain floods in the mountain valley induce rather fast changes in sediment transport rates for Orkhon river upstream (O-1, fig 2). 15-times increase of sediment load during the day was reported as a response to an intense rainfall that amounted to a total of 50 mm (29-31 July 2011), which corresponds to 20 % of the annual average precipitation.

These findings give an important conclusion that the observed long-term changes in sediment yield have been caused by the climatic drivers of water runoff shortage with associated flood decrease. Fluctuations in sediment fluxes in the given catchment are determined by changes in sediment patterns during event conditions (interseasonal variability). Even relatively small ($\Delta = 1,3$, or 30 %) decrease of annual runoff have large impact (almost 50 % decrease) on sediment yield due to river's flood response. This gives quantitative thresholds of global sediment fluxes behavior evidencing that one of the largest impacts of climate change is through changes in the overall water balance with subsequent impacts on land cover density and thus erosion rates (Knox, 1993). On the other hand, our results based on COSMO-CLM model indicate that land use changes (e.g. land disturbance by mining operations) may have an impact on water balance and could lead to the increased instability of the summer precipitation fields. Future runoff increase will be associated with growth of maximal discharges and floods magnitude and may imply then increased transport of sediments and contaminants (e.g., Chalov et al., 2014).

3.3 Importance of export of land-derived constituents and their connections with hydroclimatic conditions

Since the sediments in many cases contain relatively high metal concentrations from, e.g., mining operations (Thorslund et al., 2012), shifts in sediment transport patterns from altered hydrology within the Selenga River Basin may also influence the heavy metal loading to Lake Baikal. Above-discussed historical trends hence suggest lower loads of metals in recent years, all other conditions equal. Slight increase of total dissolved solids and twice increase of SO_4 (from 6.2-8.6 in 1950-1970 to 10.6-16.4 mg/l during 1995-2011) was caused by the lasting low-water period. Origin of metals and sediment-associated chemical constituents could be related both to natural and anthropogenic drivers. The examples of the upper Orkhon, upper Selenga and Tuul rivers, as far as downstream Selenga river represent contrasting environmental condition (table 4). Upper Selenga is relatively undisturbed part of Selenga catchment with typical distribution of pastures as the only dominated type of land use. This is forested part of Mongolia. The upper Orkhon contains few small towns and mining sites. Tuul river drains the driest and mainly steppe part of Selenga catchment, which contains both largest mining and industrial center of Mongolia and regarded to be the mostly polluted river in the region. Half of the Mongolian population is concentrated within its capital (Ulaanbaatar). Due to poor maintenance, lack of spare parts, outdated equipment, and frequent power outages, waste water from the wastewater treatment plants in Ulaanbaatar might be released directly into the Tuul River without treatment. Geochemical patterns in various parts of the catchment are determined by both natural and technogenic factors. Regional petrology causes the general enrichment of suspended matter and sediments by As, Cd, Sn, Sr, W, Pb in comparison with the lithosphere averages. The common feature of the basin consists in a prevailing transport of dissolved forms of chemical elements which a highly mobile in alkaline environment (Table 4). The share of dissolved forms of B, As, Mo, Cr, U increases (in some cases up to 98%) mostly during the low-water period (June 2012). The high water period (July 2011) is characterised by increasing turbidity of river water and the growing importance of the particulate forms of heavy metals (Fe, Mn, Ni, Co, Pb, Cu, Zn).

Concentration of chemical elements in the particulate form	Orkhon catchment		Tuul river catchment		Selenga river catchment, upstream		Selenga river catchment, downstream	
	July 2011	June 2012	July 2011	June 2012	July 2011	June 2012	July 2011	June 2012
> 75 %	V, Cr, Mn, Fe, Co, Ni, Cu, Zn, Pb	-	Mn, Fe, Co, Zn, Pb	-	V, Mn, Fe, Co, Zn, Pb	Fe	V	V, As
50-75 %	As, U	Fe, Cu	V, Ni, Cu	Fe	Ni, Cu	Cr, Mn, Co, Cu, Pb	Cr, Mn, Fe, Co, Ni, Cu, Zn, As, Pb	Mn, Fe, Co, Ni, Cu, Pb
25-50 %	B	V, Mn, Co, Ni, Pb	-	V, Mn, Co	B, As	V, Ni, Zn	-	B, Cr, Zn
< 25 %	Mo	B, Cr, Zn, As, Mo, U	B, Cr, As, Mo, U	B, Cr, Ni, Cu, Zn, As, Mo, Pb, U	Cr, Mo, U	B, As, Mo, U	B, Mo, U	Mo, U
ΔQ		2,93		2,35		1,3	1,46	
Average elevation, m	1750		1375		1750		-	
Afforestation (%)	32		32		50		-	
Population	50000		1300000		5000		2500000	
% of mining areas	0,7		0,7		< 0,1		-	

Table 4. Seasonal variability of geochemical patterns related to the hydroclimatic and anthropogenic variability (The discharge data are according to the following stations: Selenga-Mostovoy, Selenga-Zuunburen, Orkhon-Orkhon, Tuul-Ulaanbaatar)

The reported individual storm events were associated with changes in heavy metal concentrations. During a storm event reported for the Orkhon River (O1) in 29–31 July 2011, the mass concentrations (mg/kg) in the suspended load decreased for the main part of the chemical constituents during peak flow, and increased again on the falling limb, with exception of Ag and Br. Bulk concentrations (mg/l) increased during peak flow, since SPM concentrations also peaked. The only exception is As, which showed lower values during the highest discharges. The highest increase of bulk concentrations during this individual flood event was reported for Fe and Al (2.3 and 2.4 times accordingly).

References

Asselman N.E.M. Suspended sediments dynamics in large drainage basin: the River Rhine // Hydrol. Proc. 13, (1999), 1437-1450

Audry, S., Schäfer, J., Blanc, G., Bossy, C., Lavaux, G., Anthropogenic components of 550 heavy metals (Cd, Zn, Cu, Pb) budgets in the Lot-Garonne fluvial system (France). Applied 551 Geochemistry 19, (2004), 769-786.

Böhm U, Kücken M, Ahrens W, Block A and others (2006) CLM – the climate version of LM:brief description and long-term applications. COSMO Newsletter 6: 225–235

Chalov Sergey R., Jarsjö Jerker, Kasimov N., Romanchenko A., Pietron Jan, Thorslund J., Belozerova E. Spatio-temporal vari-ation of sediment transport in the Selenga River Basin, Mongolia and Russia // Environmental Earth Sciences. 73, 2 (2015), 663-680

Gellis A.C., Factors influencing storm-generated suspended-sediment concentrations and loads in four basins of contrasting land use, humid-tropical Puerto Rico // Catena,104, (2013), 39–57

Horowitz AJ. Stephens VC. The effects of land use on fluvial sediment chemistry for the conterminous U.S. - results from the first cycle of the NAWQA Program: trace and major elements, phosphorus, carbon, and sulfur. The Science of The Total Environment 400: (2008), 290-314.

Karthe, D.; Kasimov, N.; Chalov, S.; Shinkareva, G.; Malsy, M.; Menzel, L.; Theuring, P.; Hartwig, M.; Schweitzer, C.; Hofmann, J.; Priess, J. & Lychagin, M. (2014): Integrating Multi-Scale Data for the Assessment of

Water Availability and Quality in the Kharaa - Orkhon - Selenga River System. Geography, Environment, Sustainability 3(7):65-86.

Knox, J.C., 1993. Large increase in flood magnitude in response to modest changes in climate. Nature 361, 430– 432.

Кузнецов Н.Т. Гидрография рек Монгольской Народной Республики. М., Изд. АН СССР, 1955. Publication in Russian language. [Kuznetcov N.T. Hydrology of People Republic of Mongolia. Academy of sciences of USSR. 1955]

Megnounif A. , Terfous A. , Ouillon S. A graphical method to study suspended sediment dynamics during flood events in the Wadi Sebdou, NW Algeria (1973 – 2004) // Journal of Hydrology, 497 (2013), 24 – 36

McKee, B.A; Aller, R.C; Allison, M.A; Bianchi, T.S; Kineke, G.C Transport and transformation of dissolved and particulate materials on continental margins influenced by major rivers: benthic boundary layer and seabed processes. Continental Shelf Research 24 (7-8), (2004), 899–603 .

Ollivier P., Radakovitch O., Hamelin B., Major and trace element partition and fluxes 666 in the Rhône River. Chemical Geology 285 (1-4), (2011), 15-31.

Roussiez, V., Probst, A., Probst, J-L., Significance of floods in metal dynamics and export in a small agricultural catchment, Journal of Hydrology (2013), doi: http://dx.doi.org/10.1016/j.jhydrol.2013.06.013

Syvitski J. Supply and flux of sediment along hydrological pathways: research for the 21st century/ Global and Planetary Change 39 (2003) 1–11

Tananaev N.I. Hysteresis effect in the seasonal variations in the relationship between water discharge and suspended load in rivers of permafrost zone in Siberia and Far East // Water Resources 39, 6, (2012), 648-656

Törnqvist, R., Jarsjö, J., Pietroń, J., Bring, A., Rogberg, P., Asokan, S.M., and Destouni, G., Evolution of the hydro-climate system in the Lake Baikal basin. Journal of Hydrology. 519, (2014), 1953–1962

Unger-Shayesteh K, Vorogushyn S, Farinotti D, Gafurov A, Duethmann D, Mandychev A, Merz B What do we know about past changes in the water cycle of Central Asian headwaters? A review. Global Planet Change 110 (Part A), (2013) 4–25

Vanmaercke M., Zenebe A., Poesen J., Nyssen J., Verstraeten G., Deckers J. Sediment dynamics and the role of flash floods in sediment export from medium-sized catchments: a case study from the semi-arid tropical highlands in northern Ethiopia, J. Soils Sediments 10, (2010), 611-627

Walling, D.E. and Fang, D. Recent trends in the suspended sediment loads of the world's rivers // Global and Planetary Change, 39, (2003), 111-126

Zwolsman, J.J.G., van Eck, G.T.M., van der Weijden, C.H. Geochemistry of dissolved trace metals (cadmium, copper, zinc) in the Scheldt estuary, southwestern Netherlands: impact of seasonal variability. Geochimica et Cosmochimica Acta 61, (1997), 1635-1652.

Hotspot Pollution Assessment: Cities of the Selenga River Basin

Kosheleva N.[1], Kasimov N.[1], Gunin P.[2], Bazha S.[2], Enkh-Amgalan S.[3], Sorokina O.[1], Timofeev I.[1], Alekseenko A.[4], Kisselyova T.[1]

[1] Moscow State University, Faculty of Geography,
[2] Institute of ecology and evolution problems, Russian Academy of Sciences
[3] Institute of geography, Mongolian Academy of Sciences
[4] National Mineral Resources University (University of Mines), Mining Faculty

Abstract

In this study, a geochemical assessment of two large cities (Ulaanbaatar, Darkhan) and four mining towns (Erdenet, Zakamensk, Zaamar, Sharingol) in the Selenga river basin was conducted. The features of geochemical transformation of soils in the urban environments and priority pollutant sources were revealed. The detailed study of concentrations of heavy metals in the urban soils of the selected sites, characterized by different land-use types, showed that the associations of accumulating heavy metals are determined by the emissions of industrial enterprises and motor vehicles, while the levels of heavy metals are controlled by soil properties. The geochemical conditions in the mining centers are derived from geochemical features of parent rocks and ore bodies, which are usually rich in heavy metals. The environmental risks associated with the contamination of the mining landscapes are determined not only by the high background concentrations of ore and ore-linked elements in soils and rocks, but also by the technologies used in the mining operations, and by the volumes of the produced waste.

Introduction

Development of industrial cities and rapid growth of urban population in the Selenga basin causes multiple ecological problems in the region. The manmade pressure on the environment is accompanied by air and water contamination, accumulation of pollutants in soil cover, bottom sediments and vegetation. Besides deteriorating environmental conditions, the contamination also affect public health, threaten biodiversity and put at risk the prospects for sustainable development both on the spot and also in a wider area. The purpose of the present study is to evaluate the pollution of the environment with heavy metals (HMs) in the largest industrial centers located in the Selenga river basin. The investigation is comprised of the following steps:

- geochemical study of non-polluted (background) territories;
- identification of priority pollutants and evaluation of geochemical transformation of the urban environments using the geochemical background values;
- environmental risk assessment on the basis of maximum permissible concentrations (MPC) specified in Soil Quality Standards of Mongolia and Russia.

The studies considered the following objects: 2 large cities in Mongolia (Ulaanbaatar, Darkhan) and 4 mining towns. Among them are Erdenet and Zakamensk, which are known as the largest centers of non-ferrous metallic mineral mining; Zaamar, located in the Tuul river valley, with large-scale placer gold mining operations and Sharingol which is the place where sulfur coal is produced in open-cast mine. The studies began in 2008 and are still in the process: in 2014 they included the territory of Ulan-Ude.

Study area

The study area is characterized by complex mountain, hollow and valley relief; the cities are situated in the valleys of the Selenga tributaries. The climate is extremely continental. Soil-forming rocks are represented mainly by alluvial and proluvial deposits and also by shales and granitoids. Forests on mountain-forest soils cover slopes with northern aspects; dry shrub- and meadow-steppe vegetation on mountainous chernozems and chestnut soils occupies south-facing slopes (Nogina, 1984).

The cities differ significantly from each other in terms of area, population and the number of motor vehicles. The population of Ulaanbaatar exceeds one million people. The common sources of pollution in all cities are power plants and ger districts. These entities use brown coal which combustion products pollute the atmosphere. Different sets of elements in ash emitted by the power plants and by coal stoves in private houses originate due to different temperature of combustion. Other major sources of pollution are associated with various industries, motor vehicle fleet using leaded gasoline in Mongolia, and terrigenous dust in the south of the region

Materials and methods

The samples of the urban soils, surface technogenic materials, collected in the tailing areas and at landfill sites, the sediments, accumulated on the bottom of river channels and ponds and samples of brown coals and fly ash collected in Ulaanbaatar power station (Table 1) were examined for total contents of 20 HMs and metalloids by mass-spectral and atomic-emission methods with inductively coupled plasma (ICP-MS) at the All-Russia Research Institute of Mineral Raw Materials (Moscow). Additionally, in some cities the samples of vegetation, snow solutions and suspended load were also analyzed for HMs content.

The samples of the urban soils were collected from the upper horizons (0-10 cm) using a regular grid with relatively even spacing in all functional zones of the cities. Background conditions were studied within recreational areas or in natural landscapes adjacent to urban territories.

Geomedia	Ulaan-baatar	Ulan-Ude	Dark-han	Erde-net	Zaka-mensk	Zaamar	Sharyn-gol	Total
Topsoil	90	293	155	260	141	79	82	1100
Soil pits	18	–	3	23	35	50	4	133
Landfill & tailing material	–	–	–	10	14	–	8	32
Bottom sediments	5	5	2	10	11	11	12	56

Table 1. Number of samples collected in the cities and mining centers

The extent of soil pollution was determined in relation to reference (background) sites using enrichment factor and integrated soil pollution index. The enrichment factor (EF) was calculated for the each element and for the each sampling site using the formula EF=Ca/Cr (Ca>Cr), where Ca is the measured concentration of each metal or metalloid in urban soils and Cr is the reference concentration of the element in the background soils. The cumulative techno-genic impact on urban soils, related to multielement pollution, was described on the basis of the integrated pollution index $Zc=\Sigma EF - (n-1)$, where n – number of elements with EF >1.5 (Environmental geochemistry, 1990).

The environmental risks associated with landscape pollution are deter-mined on the basis of the integrated pollution indices (Table 2) for the whole territories of the cities, as well as for their individual land-use zones and also for each sampling site. The maps displaying the distributions of these indices help to identify spatial patterns in contamination of the urban territories.

Pollution levels / environmental risks	Dust fall-out P_n, kg/km² per day	Integrated pollution indices		
		Heavy metal immission Z_d	Snow pollu-tion Z_c	Soil pollu-tion Z_c
Low /Acceptable	< 200	< 1000	< 32	<16
Moderate / Moderate	200 – 300	1000 – 2000	32 – 64	16–32
High / High	300 – 500	2000 – 4000	64 – 128	32–64
Very high / Very high	500 – 800	4000 – 8000	128 – 256	64–128
Extremely high / Ex-tremely high	> 800	> 8000	> 256	>128

Table 2. The levels of contamination of soils and snow cover in urban landscapes and the corre-spondent degrees of associated environmental risks (Kasimov et al., 2012)

Statistical treatment of the data was performed using Statistica 7, for the cal-culations of the indices MS-Excel was used. Soil and geochemical data were visualized and geochemical maps were constructed using ArcGIS 10 and MapInfo 12. To reveal the major factors and conditions which may control ge-ochemical heterogeneity of the urban areas, multivariate regression analysis was performed using SPLUS package.

Results and discussion

The assessment of the pollution in the large cities

Ulaanbaatar, the capital of Mongolia, is the most particulate matter-polluted city in the world (Gutticunda, 2007; http://www.nso.mn/v3/). Its population is 1.2 million persons. One of the main sources of pollution is fuel-energy complex. It includes three power plants and households in ger areas. In the power plants and coal-fired ger stoves the brown coals from Nalayh, Baga-Nur and Sharyn-gol mines are used. Relative to the composition of the continental crust, the burned coals contain higher amounts of Pb, As and Mo, and also Cu, Sr, Cd, Ni. Compared to other brown coals of the world, they are rich in Pb, Cu, Ni, W and Mo. Another source of pollution is motor vehicle fleet represented by 300 thousand vehicles that use leaded gasoline with a high content of Pb. Besides, the construction, textile and food processing industries are located in Ulaanbaatar and serve as additional pollution sources.

The comparison between the background concentrations of heavy metals and metalloids in the soils of the study area and the global abundances of the elements in the lithosphere (the elements' clarkes) showed that the regional geochemical background conditions are associated with high levels of As and Cd and low concentrations of other HMs. Relative to the reference sites, the urban soils, show the high degree of accumulation of Pb (EF=3.1) and Zn (EF=2.6). The mean values of EF for other trace elements do not exceed 2. The intensity of accumulation is decreasing in order: Mo> Cr> Cu> Cd> Ni> Co> Sr> V> As. The most stable association traced in the soils of Ulaanbaatar, are Ni-Co-V, As-Sr and Cu-Cd-Cr-Zn. The first association is derived from the composition of the parent rocks (shale and clay), the second association combines the elements traced in ash emissions from coal combustion and the third one includes the elements from the anthropogenic sources such as emissions from industrial enterprises, motor vehicles and domestic waste.

The geochemical specialization associated with the functional identity of the territories is manifested in noticeable pollution of soils in the industrial zone of Ulaanbaatar with Zn, Mo, Cr, Cd, Pb, Cu (EF > 2.0). The soils in the residential area with many-storeyed apartments were found to be more intensively contaminated, especially with Pb (EF=5.8) and Zn (EF=3.2), than the soils in the areas with one-storeyed housings. The accumulation of HMs is more pronounced in the central part of the city, where a geochemical anomaly is formed due to the impact of many different sources of soil pollution (residential, industrial and traffic) with common element associations: Ni-Co-V; Cu-Cd-Zn; As-Mo.

The spatial distributions of HMs in the urban soils are determined by the sources of their origin: Zn, Pb, Cd, Cu are introduced to the urban environment mainly through transport emissions, As, Sr, V are released by the fuel energy sector, the additions of Ni, Co, Cr are associated with the construction industry. Mo, that exists in anionic form, can be transferred to and accumulated in the soils of subordinate positions. The major spatial patterns in HMs distribution are complicated by numerous local technogenic anomalies.

The concentrations of the majority of the studied heavy metals are controlled by soil properties: humus content (As, Cd, Cu, Mo, Zn), proportion of clay (Ni, Co), levels of sulfates (Pb, Sr), pH (Cr). However, the variations in V concentrations are found to be dependent on the type of anthropogenic impact which varies with functional identities of territories.

The values of the integrated pollution index Zc in the industrial area, in the city's center and around the major city's highways reach 16–32 units (Fig. 1a). The least pollution with Zc < 8 is observed in the western part of the city. The anomalies in the residential area with private buildings are characterized by Zc values less than 30 units. They are formed due to the additions of HMs through coal ash release and due to emissions from traffic and domestic landfills. In the vicinity of sewage treatment plants the anomalies (Zc=26) are related to the sewage sludge impact. The environmental risks associated with the pollution are characterized by the percentage of the area where the concentrations of HMs in soils are above the established sanitary standards (the MPC). It was found that 100 % of the Ulaanbaatar territory has higher levels of As; 38 % of the city's area have high levels of Zn; 20 % − of Mo; 18 % – of Pb and 4 % – of Cr.

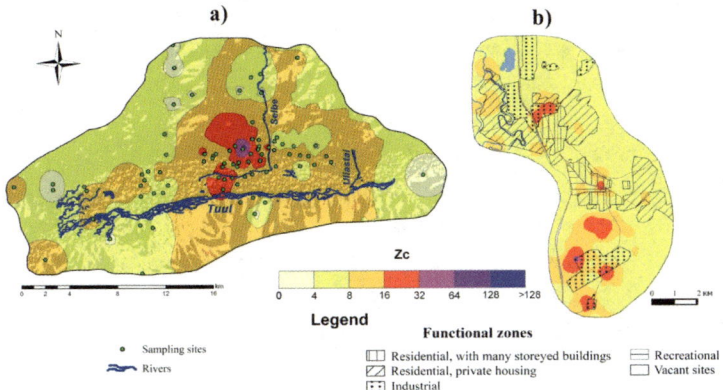

Fig.1 The distribution of the integrated soil pollution index Zc in Ulaanbaatar (a) and Darkhan (b)

Darkhan, with a population of 93 thousand, is the third largest city in Mongolia and a major industrial center with a cement, a leather tanning, an iron-steel plants, a power plant and some other industrial enterprises. The residential area with modern apartments where 86 % of the citizens live occupies the central part of the city. The ger districts are located in the eastern, northern and western outskirts of the city. The transport network is represented by major highways with asphalt cover and the railway. A significant part of the city is occupied by vacant territories.

Regional soil geochemical background conditions are manifested by high contents of W, As, Cd, Sn, Pb and low contents of Co, Cr, Cu, Ni relatively to clarkes (Grigoriev, 2009; Taylor, McLennan, 1985; Turekian, Wedepohl, 1961; Vinogradov, 1959). The urban soils accumulate Sb, Pb, W, Mo, Cu, Cr, Zn, As, Cd, however the difference between the observed concentrations and the background values is not significant (EF=1.75-1.25). Even so, each land-use zone is characterized by its own set of accumulating elements.

In the industrial zone the soils are enriched with Pb, Sb, Zn, W, Cd, Cr (EF=3.2-1.8). A priority pollutants in the traffic network are Cd (EF=2.9) and also Cu and Sb. The soils in the residential area with many-storeyed apartments are characterized by high concentrations of Sb, Cd, Zn, Pb, Cu and Cr (EF=2.2-1.3), the soils in the area with one-storeyed buildings are rich in Sb,

Zn, Pb, Cu, Cr, Cd (EF=1.7-1.25). The soils of the recreational area are characterized by high levels of W (EF=3.7), and those of the vacant territories have high concentrations of W, Mo, Sb, As and Cr (EF=2.5-1.5).

The paragenetic associations of the elements traced in the urban soils of Darkhan are As-Mo-W, Zn-Cd-Cu, Co-V-Cr-Ni and Sb-Pb. The first association combines the elements added to soils due to burning of brown coal in the power plant and in individual stoves (Sorokina et al., 2013), the second association is derived from techngenesis, the third one combines the elements that exist and migrate in cationic form.

The levels of HMs accumulation are defined mainly by soil properties. For the majority of the heavy metals (Zn, W, Cd, Mo, Cr, Ni, V) and also for the integrated soil pollution index Zc the main controlling factor is the content of Fe oxides in the soil. The accumulation of Cu and Sn is controlled by humus content, pH values influence the accumulation of Pb and the content of Mn oxides defines the distribution of As.

The integrated soil pollution index Zc estimated for Darkhan is 7, which indicates low contamination level (Fig. 1b). The anomaly with Zc > 32 is restricted to the industrial zone in the south of the city and occupies the adjacent vacant territory which acts as a buffe zone, protecting the residential areas from the emissions of industrial enterprises. Another anomaly is localized near the grinding plant and along major highways.

The excess over the MPC is revealed for Pb, Mo, As, Zn, Cu, Ni, Cr and V. The percentage of the area with the excessive concentrations of certain elements is diminishing in the order: As (97 %) > V (26 %) > Mo (17 %) > Pb (9 %) > Cr (4 %).

The assessment of the mining centers

Erdenet. Erdenet is a mining town where since 1976 one of the largest world deposits of copper and molybdenum ore has been being exploited. Currently 27.8 million tons of ore per year is mined. As a result of 40-year history of the mine operation a huge tailing area covering more than 1500 hectares has been formed.

There is a natural geochemical anomaly in the soils formed on Permian volcanic and volcano-sedimentary deposits which is manifested in high concentrations of Se, Mo, Sb, Sr relatively to clarkes of the lithosphere.

The list of priority elements polluting the soils in all land-use zones of the city, except for the ger area, includes ore elements (Mo, Cu) and the heavy metals released by burning gasoline, coal and domestic waste (Bi, Cd, Pb, Zn) (Fig. 1). The highest technogenic load is observed in the industrial zone where the accumulation of Mo (EF=10.7), Cu (EF=10.6), Se (EF=2.3), As (EF=1.6), Sb (EF=1.5) and W (EF=1.5) occurs. The minimum loading is observed in the ger areas at the furthest locations on the leeward side of the ore deposit. The elements with similar spatial distributions, notably with common areas of accumulation and dispersion in the surface layer of the urban soils, are incorporated into three stable paragenetic association reflecting compositional features of the parent rocks: Cu-Mo-As-Sb-Bi, Zn-Cd-Pb-Sn and V-Co-Cr-Ni.

The elements of the first association form two distinct anomalies: one in the tailing zone and another within the territory of the ore processing factory "Erdenet", where the highest levels of contamination occur and the concentrations of Cu-Mo-As-Sb-Bi-Cd-Zn are respectively 982, 460, 44, 42, 12.6, 10.7 and 5 times higher than in the reference samples. The second association forms two less distinct anomalies in the residential area with many-storeyed apartments and on the territory of the city's landfill. The third association includes the elements which average contents have rather homogeneous spatial distribution and are very close to the reference values due to the low contents of these elements in porphyry complexes typical for the ore field.

The spatial patterns of Cu, Mo, As, Sb, Zn, Cd and Pb distribution are determined by the degree of technogenic disturbances. The spatial distribution of other elements – Ni, Bi, Ba, W, Se, V, Co, Sn, Cr, Cs, Sr – are controlled mainly by soil properties. The concentrations of the elements which belong to the first group are 1.4-5 times higher in the industrial and residential areas than in other functional zones. Within some areas the distribution patterns of the elements belonging to the first association are controlled by soil properties: pH values (Cd, Pb), humus content (Zn), the contents of Fe and Mn oxides (As, Cu, Mo, As, Zn) and soil texture (Cd). The second group of HMs and metalloids is accumulated together with Fe oxides (Ni, Bi, Ba, W and Se); Mn oxides (V,

Co and Sn); Al oxides (Cr and Cs). The trace elements are fixed in neoformations and various minerals.

The highest increase (3.8-8.6 times) over sanitary standards of Mongolia are revealed for As, Cu and Mo in the industrial zone (Fig. 2). The analysis of the Zc values showed that half of the city can be considered as clean (Zc <8), and 16% of its territory − as slightly polluted (Zc=8-16). The eastern part of the city, located within the industrial zone, is characterized by very high levels of soil contamination (Zc=64-128). Distinct anomalies are restricted to the tailings and to the territories of ore processing plant "Erdenet" and a smaller plant, called "Erdmin"" (with maximum Zc =1560).

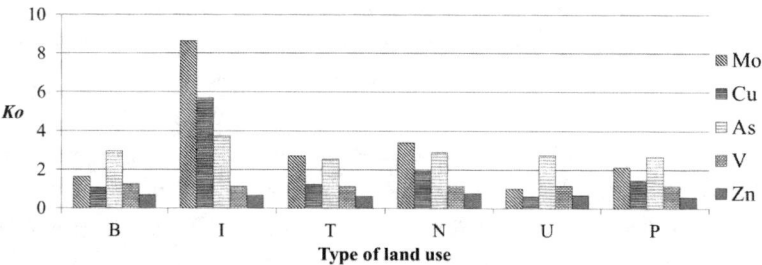

Fig. 2. The coefficient *Ko*, showing the excess of Mo, Cu, As, V, Zn concentrations in soils of Erdenet in various functional zones over the established standards (the MPC). Zones: *B − background areas; I − industrial zone; T − transport network zone; N −residential area with many-storeyed apartments; U − ger areas; P − grasslands.*

Zakamensk. The primary industry in Zakamensk is mining of Mo, W, and Au reserves. The ore processing plant has produced 44.5 million tons of waste which were stored in the Dzhidinsky, Barun-Naryn tailing areas and also in the emergency tailing basin. In 2011 nearly 3.5 million tons of waste were translocated to the top of the Barun-Naryn basin from the emergency tailing.

Local geochemical background conditions in soils, relatively to clarkes in the Earth's crust (Grigoriev, 2009; Taylor, McLennan, 1985; Turekian, Wedepohl, 1961; Vinogradov, 1959), are characterized by high contents of W, Mo, Bi, Sb. The first-priority pollutants for the urban soils are W, Bi, Cd, Pb and Mo (Fig. 3). The highest technogenic loadings were identified for the soils in the industrial zone and in the residential area with many-storeyed apartments. The list of contaminants in the industrial zone includes Bi (EF=23.6), W (EF=21), Cd (EF=10.8), Be (EF=8.1), Pb (EF=8.0), Mo (EF=6.9), Sb (EF=6.6).

In the residential area the list of contaminants is represented by W (EF=6.0), Bi (EF=5.2), Cd (EF=4.8), Pb (EF=2.6), Be (EF=2.5), Zn (EF=2.4) and Cu (EF=1.5).

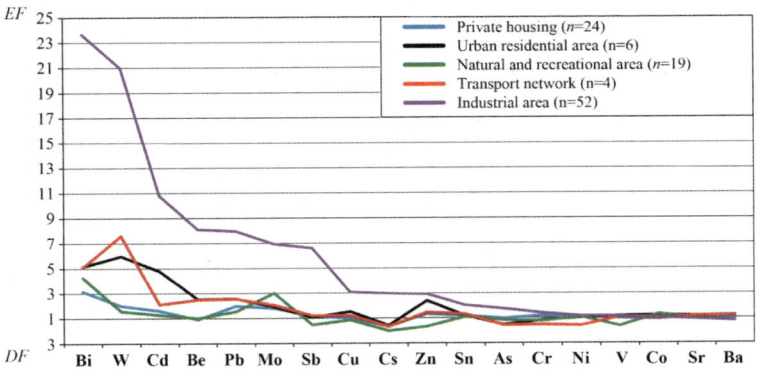

Fig.3 Geochemical spider diagrams for the soils located in different functional zones of Zakamensk

Two stable associations are identified in the surface horizons of the urban soils: Be-Cs-W-Bi-Cu-Zn-Cd-Sn and As-Sb-Pb-Mo. They combine mainly chalcophile elements which are very typical for the rocks of Pervomaiskii stockwork, Inkurskoye and Kholtoson deposits. Both of these two elements' associations have a common accumulation zone in the city's centre where the tailing areas are located. The concentrations of W are 529-824 times higher than the element average concentration in uncontaminated world soil (Kabata-Pendias, 2011) the concentrations of Sb are 91-657 times higher, Mo – 23-255 times, Bi – 83-188 times, Cd – 32-80 times, Pb – 48-56 times, Cu – 9.3-46.3 times and Zn – 13-20 times higher. The maximum concentrations are restricted to Barun-Naryn sluge pond. They are associated with buried humus horizons, which were exposed to the day surface as a result of remediation operations.

The spatial distribution of the integrated soil pollution index Zc is determined by the nature and the intensity of anthropogenic impact. In the industrial area, relatively to reference sites, the distinct anomalies (Zc=185-710) are formed in technogenic surface materials with low clay content and acid reaction. Acid reaction is typical for the tailings areas, disposal channel and Modonkulskoe technogenic mineral deposit. In other areas the variations of Zc values are determined by natural landscape-geochemical processes and factors:

heavy metals are accumulated in the soils of subordinate positions at biogeo-chemical barriers. The capacities of the barriers are related to the amount of humus in soils. The accumulation positions (with average values of Zc=36.7) are restricted to the floodplain and the terrace of the Modonkul river occupied by private housing and natural recreational landscapes.

The greatest environmental risk is posed by Pb, Sb, Cd and As, which concentrations in the soils of the industrial zone are 1.7-7.8 times higher than the MPC.

Zaamar. The development of the Zaamar gold field started in 1994. In the pro-cess of gold recovery dragline excavators and dredges are used. There is a secondary hydraulic mining from the waste rock done by Illegal miners. The official gold mining area covers more than 4000 hectares and together with the surrounding territories represents 5 main land use units. They are the land-scapes of current placer gold mining, the landscapes used previously for waste rock disposal and hydraulic tailings, remediated territories, aquatic landscapes, weakly disturbed or undisturbed landscapes. Due to excavation works on the floodplain the channels for dredges are created. The erosion of the ponds' and channels' dikes contributes to high sediment load in the river Tuul. Geochemi-cal changes in the man-made environment are defined by the method of gold mining which involves the settling of particles with high specific gravity and re-moval of other materials to the drainage system. As a result, a sharp rise in suspended load in the river Tuul and the changes in chemical composition of alluvial sediments caused by changes in sediment granulometry and mineral-ogy are observed.

In technogenic surface materials and in the soils of the study area the following association Co-As-W-Ni-Zn-Sr is accumulated, providing a number of local maxima in the area of current gold mining (EF up to 2.8). High concentra-tions of Zn, Ni, Sr occur more frequently. The maximum level of Zn and Ni is typical for the areas of current mining; while in the areas of previous waste rock disposal and particularly at the remediated sites the levels of the elements drop and even become lower than the background concentrations, except for Sr, characterized by very high concentrations.

The major factor controlling the accumulation of As, Co, Cu, Mo, W in technogenic surface materials and in the soils of the study area is the content of Fe oxides serving as sinks for a wide spectrum of metals. The factor of secondary importance is pH. The rise of pH values from 7.3 to 9.9 increases the accumulation of As, Co, Cu, Mo in the surface soil horizons and also in the ground. Mn oxides contribute to the accumulation of As and W due to their binding under oxidizing conditions into stable organo-mineral compounds. The soils of the undisturbed areas are distinguished by low concentrations of As, Co, Cu, and W. Mo has low concentrations in old waste rock piles. The accumulation of Mo and W is controlled by humus content that varies between 0.5-3.4 % in the upper (0-10 cm) soil horizons. The location of the soil within the catena only affects the content of Co: the maximum of Co concentrations are commonly found at the autonomous (summit) positions.

The assessment of the aquatic landscapes was performed on the basis of the chemical analysis of bottom sediments and suspended material sampled in the river with grain-size of 0.05-0.1 mm. Within the mining area the contents of Cu, Ni, Co, As, Pb in the bottom sediments increase downstream. The maximum concentrations of Cd, Bi, Cu, Ni, Co, As, Pb, Sn, Sb, Mo, Cr, Zn, V, Ba (EF=1.5-3.0) are restricted to the depositional geochemical barriers which occur in the ponds left in the valley after the works of dredges. The pollution of the ponds, which are enclosed systems, is the result of two technogenic factors – official mining and secondary gold recovery by illegal miners. The bottom sediments of the Tuul river are rich in Sb, Sn, Pb, compared to the global clarkes of the lithosphere (Grigoriev, 2009; Taylor, McLennan, 1985; Turekian, Wedepohl, 1961; Vinogradov, 1959).

The main sources of the suspended load in the territory of placer gold mining are the eroding overburden and the tailings produced by dredging. The maximum concentrations of 13 trace elements in the suspended sediments (with EF=1.5-2.0) are restricted to the territory of placer gold deposit. Downstream, to the confluence of the Tuul river with the Orkhon river, a gradual decrease in the elements' concentrations along 50-kilometer stretch is observed.

The concentration of HMs in soils, technogenic surface materials, bottom sediments and suspended load were compared with the MPC, specified in Mongolian Soil Quality Standards. The high ratio between the observed elements' concentrations and their established standards in surface materials and soils of the tailing and remediated areas was found only for As and V. On average, the concentrations of As are 3.7 times higher than the established standards and at certain sites they are 9.3 times higher. The concentrations of V are 1.1 - 1.5 higher than their standards. In the background landscapes the excess of As concentrations over the MPC is very common and the difference between the observed levels and the established standards is 2 times and more. The excess of Sr (1.2 times) was found only at certain reference sites. In the sediments accumulated within the mining area on the bottom of the Tuul river and the ponds the concentrations of Hg are 4.7 times higher than the established standards and the concentrations of As 2.9 times (on average) or 6.5 times higher (at certain sites). The contamination of the bottom sediments with Hg is the result of amalgamation process used during illegal gold mining. The concentrations of V are very close to the established standards and for 8 trace elements they are even lower than the MPC. On average, the difference between the observed concentrations and the established standards is 1.7-8.0 times.

Sharyngol. In the town with a population of over 10 thousand the brown coal mining was conducted up until 1964. The Sharyngol open cut coal mine is 1.5 km long and 240 m deep. As a result of 50-year history of the mine operations, the waste tips reaching 30 m and covering nearly 800 hectares (which is twice as large as the town's area) have been formed. Mineralogical and petrographic composition of the overburden varies considerably due to the mixing of Quaternary deposits with deeper strata. The most significant emissions from coal combustion are recorded in winter period resulting in contamination of the urban landscapes. In the study area 7 landscape and land-use units are identified: the residential areas with many-storeyed apartments and low-rise buildings, the waste tip areas, the placer gold mining sites along floodplain sections of the Sharyngol river, vacant territories and landfills, arable lands, weakly disturbed and undisturbed landscapes.

The surface technogenic materials of the overburden are characterized by high contents of W (EF=4.9), As (EF=3.4), Mo (EF=2.7), Bi (EF=1.8) (Fig. 4). The urban soils located within the residential area with many-storeyed apartments accumulate Pb (EF=1.8), W (EF=1.6), Zn (EF=1.5). In the residential area with private low-rise housing the soils are relatively rich in W (EF=1.8),

Pb (EF=1.7), Mo (EF=1.6), Sb (EF=1.5). The geochemical anomalies formed in the soils of Sharyngol due to the deflation of overburden are overlain by the anomalies with similar association of elements derived from brown coal fly ash. A significant addition of Pb and Zn to the urban soils is found to be due motor vehicles' emission (Sorokina et al., 2013).

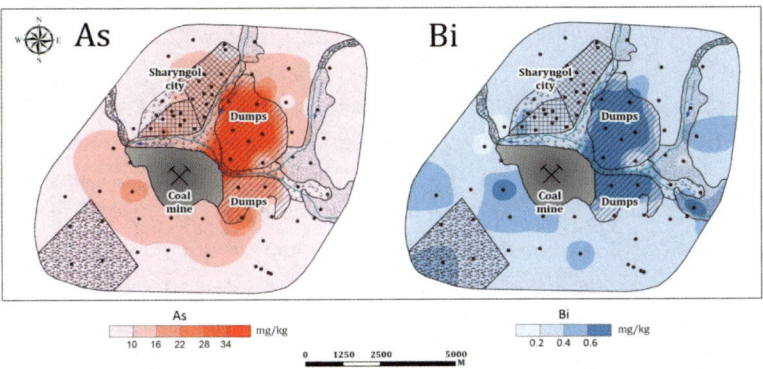

Fig.4. Geochemical anomalies in the areas of waste tips (dumps), formed during opencast brown coal mining

The bottom sediments of the Sharyngol river in the urban environment are characterized by elevated concentrations of Mo and W (EF=5.9), As (EF=5.0), Cu (EF=2.5), Bi (EF=1.4). The additions of the elements are derived from the discharges of mining waste water and water from municipal sewer treatment system. The pH values in the sediments decrease sharply from 8.2 to 7.1, which is due to the discharges of formation water pumped out during the coal mining. In the vicinity of the open cut mine the discharge of suspended load increases from 7.7 to 37 g/s. The maximum concentrations of As, Cu, Mo, Ni, W, Sb (EF=1.5-2) in the suspended load are found within the area of the coal deposit. However, downstream, at the distance of 80 km just before the con-fluence with the Orkhon river, a sharp decline in the elements' concentrations is observed.

The leading factor, controlling the accumulation of As, Bi, Mo, Pb, W, Zn in the urban soils and surface technogenic materials, is the affiliation of the territory with the waste tips. The growth of pH values promotes the loss of As,

Mo, Pb, W, Zn and the accumulation of Bi; the increase of humus content contributes to the accumulation of all the elements, except for As. In the soils, occupying footslopes and the floodplain, all the elements, except for Mo and W, are accumulated. The increase of Fe oxide content is associated with the losses of As, Bi and Zn, while the accumulation of Mn oxides is accompanied by the deposition of Bi and W and the loss of Mo.

The environmental risks of pollution associated with the waste tips are defined by high levels of As (the concentrations of As are 8.3 times higher and in some cases 18 times higher than the established standards), Mo (1.3-5.7), V (1.1-1.5). For As the excess of the observed concentrations over the MPC was recorded in 100% cases, V – in 67 % cases, Mo – in 25 % cases. The local maxima of the integrated soil pollution index Zc in the area of the waste tips varies from 36.1 (high level of environmental risk) to 75,3 (very high level) however on average it equals 22.5 indicating the moderate level of environmental risk. In the urban soils the concentrations of As 2.9 times higher than the established standards, Mo – 1,1 to 1.4 times higher and Zn −1.1 to 3.4 times higher. The integrated soil pollution index Zc on average equals 11.3, indicating the acceptable level of environmental risk however the local maxima of Zc vary from 17.0 to 21.3. The concentrations of Cu, Mo, Ni, Zn in the water of the Sharyngol river are 1.3-19 times higher than the standards established in fishery.

Conclusions

The studied cities can be viewed as pollution hotspots: they are the areas which receive high pollution loads and pose a threat to the surrounding environment and public health. The associations of heavy metals, accumulating in the urban soils, are determined by the emissions of industrial enterprises and motor vehicles, however the levels of heavy metals are controlled by the properties of urban soils, defining their absorption capacity. If the granulometry of the urban soils is dominated by coarser fractions and the soils are characterized by low humus content, the significant portion of the pollutants does not accumulate in the soil medium, heavy metals can be involved into biogeochemical cycling and released to the surrounding ecosystems through water and air fluxes.

The geochemical conditions in the mining centers are caused by geochemical features of parent rocks and ore bodies, which are usually rich in HMs. The environmental risks associated with the contamination of the mining land-scapes are determined not only by the affiliation of the areas with metallogenic anomalies with high concentrations of ore and ore-linked elements, but also are influenced by the methods and technologies used in the mining operations, as well as by the volumes of the produced wastes. The location of waste and tailings is important regarding public health: the tailing areas should be located far from the residential areas in accordance with the wind rose, as well as they should be protected from water and wind erosion.

References

Геохимия окружающей среды / Сает Ю.Е., Ревич Б.А., Янин Е.П. и др. М.: Недра, 1990. 335 с. [Environmental geochemistry, 1990. Saet Yu. E., Revich B.A., Yanin Ye.P. et al. (eds.). Nedra, Moscow, 335 p. (In Russian)]

Григорьев Н.А. Распределение химических элементов в верхней части континентальной коры. Екатеринбург: УрО РАН, 2009, 383 с. [Grigoriev N.A. (2009) Chemical element distribution in the upper conti-nental crust. UB RAS, Ekaterinburg, 383 p.]

Касимов Н.С., Битюкова В.Р., Кислов А.В., Кошелева Н.Е., Никифорова Е.М., Малхазова С.М., Шартова Н.В. Проблемы экогеохимии круп-ных городов // Охрана и разведка недр, 2012, № 7, с. 8-13. [Kasimov N.S., Bityukova V.R., Kislov A.V., Kosheleva N.E., Nikiforova E.M., Malkhazova S.M., Shartova N.V. (2012) The problems of the environ-mental geochemistry in large cities. Mineral protection and prospecting 7, 8-13. (In Russian)]

Ногина Н.А. Почвенный покров и почвы Монголии. 1984. 192 с. [Nogina N.A., 1984. Soil cover and soils of Mongolia, 192 p.]

Сорокина О.И., Кошелева Н.Е., Касимов Н.С., Голованов Д.Л., Бажа С.Н., Доржготов Д., Энх-Амгалан С. Тяжелые металлы в воздухе и снеж-ном покрове Улан-Батора // География и природные ресурсы. 2013.

№ 3. C. 159-170. [Sorokina O.I., Kosheleva N.E., Kasimov N.S., Golovanov D.L., Bazha S.N., Dorjgotov D., Enkh-Amgalan S. (2013) Heavy metals in the air and in the snow cover of Ulan-Ude. Geography and natural resources 3,159-170.]

http://www.nso.mn/v3/ (The Official Website of National Statistical Office of Mongolia)

Gutticunda S. (2007) Urban air pollution analysis for Ulaanbaatar. The World Bank Consultant Report. Washington DC, USA,125 p.

Kabata-Pendias A. (2011) Trace elements in soils and plants. 4th ed. CRC Press, 548 p.

Taylor S.R., McLennan S.M. (1985) The continental crust: Its composition and evolution. Blackwell Scientific Publishers, Oxford, 312 p.

Turekian K.K., Wedepohl K.H. (1961) Distribution of the elements in some major units of the Earth's crust. Geological Society of America Bulletin 72, 175–192.

Vinogradov A.P. (1959) The geochemistry of rare and dispersed chemical elements in soils. 2nd ed. Consultants Bureau Enterprises, New York, 209 p.

Geochemical Transformation of Soils Caused by Non-Ferric Ore Mining in the Selenga River Basin (Case Study of Zakamensk)

Ivan V. Timofeev[1]

[1] Department of Landscape Geochemistry and Soil Geography, the Faculty of Geography, Lomonosov Moscow State University

Abstract

The results of the geochemical survey conducted in 2012 made it possible to identify spatial distribution patterns and abundances of 18 hazardous heavy metals and metalloids in the soils of the town of Zakamensk. The chalcophylic elements (W, Bi, Cd, Pb and Mo) appeared to be the priority pollutants nearly in all functional zones. The maximum accumulation of Bi, W, Cd, Be, Pb, Mo, Sb is restricted to the industrial area where total pollution index of soils (Z_c) more 128. Environmental assessment of surface soil horizons geochemistry in Zakamensk showed that two-thirds of its area have dangerous and extremely dangerous levels of soil pollution.

Introduction

Ore mining has a large-scale environmental impact and leads to the irreversible changes of soils, water, vegetation and atmosphere. Mining landscapes are specific technogenic landscape-geochemical systems with abnormally high concentrations of elements, mainly landscape-affecting (Kasimov, 2013). Imposing of technogenic auras and dispersion flows on natural geochemical anomalies quite often creates serious threat to the health of population. Therefore the environmental effects of mining activities are intensively studied in many countries of the world. However, the approaches to environmental assessment and zoning of litologically and geochemically heterogeneous mining areas are the least developed (Avessalomova, 1986, 2004; Makhinova et al., 2014). The principal processes and mechanisms of heavy metals (HM) and metalloids fixation in soils that complicates soil remediation and reclamation are also studied insufficiently.

The purpose of this work is to determine concentrations of heavy metals (HMs) and metalloids in soil cover of the town of Zakamensk where the Dzhidinsky tungsten-molibdenum mining plant is located, and to estimate the environmental hazard associated with soil pollution.

Study object

Natural settings. The territory of Zakamensk (45 sq.km) is located in the southern part of the Mongol-Siberian mountain belt and belongs to the Selenga-Vitim zone with moderate neotectonic activation of the earth's crust (Ufimtsev, 1991). It is situated within the border of two regional geological structures, i.e. carbonate and terrigenous Lower Paleozoic strata of the Dzhidinsky synclinorium and granitoid intrusions of the Modonkulsky massif with relative elevations of 300-400 m.

The region has severely continental climate with cold (-49°C) and long, rather low-snow winter and short warm (mean July temperature is +15,6°C) summer. The annual amount of precipitation is 250-400 mm; western and southwestern winds prevail in the area (36% and 16% respectively) (Fig. 1).

Mountain sod-taiga soils are developed in the autonomous landscapes; steep slopes with periodically washing water regime have sod-forest lithogenic soils with shortened vertical profile. Sod-forest or gray forest soils are developed on the flat lower parts of slopes and in depressions (Nogina, 1964), while the Modonkul and Dzhida river valleys have alluvial-meadow soils.

The natural vegetation includes forest-steppe and forest altitudinal belts. The flat-leaved birch (*Bétula platyphýlla*) and Siberian larch (*Lárix sibírica*) in combination with meadow steppes prevail in the forest-steppe belt; boreal light-coniferous forests with Siberian larch (*Lárix sibírica*) as a cenosis-forming species are widespread in the forest belt.

The main water course is the Modonkul River, the right tributary of the Dzhida River. The valley between watersheds is 3 to 7 km wide, the valley sides are mostly steep with erosion cutting up to 2,0-2,5 m. The main inflows – the Barun-Naryn River, the Zun-Naryn River, the Inkur stream, etc. – don't function today: tailing ponds were created on the first two of them, while the last one is turned into the cascade of reservoirs used for the extraction of placer gold and tungsten.

Technogenic impact. During 1934-2001 the town's primary enterprise was the Dzhidinsky tungsten-molibdenum mining plant (DTMMP) with a number of mining fields: stockwork molybdenum (Pervomayskoye), sulphide-tungsten (Inkurskoye and Holtosonskoye, both ore and placer) and placer gold (Myrgensheno, Ivanovka). Ores contain admixture elements of hazard categories I to III (Pb, Zn, F, Mo, W, Be, Bi, As, etc.). During the DTMMP operation 44.5 million tons of wastes were stored in Dzhidinsky (piled), Barun-Narynsky (hydraulic-mine dump) and emergency tailing ponds (Fig. 1). In the course of 2011 reclamation 3.5 million tons of waste was moved from the last one to the top part of the Barun-Narynsky tailing pond.

After acquisition of Barun-Narynsky and Dzhidinsky tailing ponds the Zakamensk JSC has evaluated the reserves of metals in technogenic sands and in 2010 started their processing with formation of a new tailing pond in the Zun-Naryn River valley.

Besides mining, wood production and processing, cast iron, steel and bronze production, construction, stone working and production of food are characteristic of the town.

Functional zoning of the city. Basing on the land use type and the level of technogenic loading the territory of the town of Zakamensk was subdivided into several functional zones (Fig. 1): two residential (private housing and urban), transport, natural-recreational (natural forests, urban parks and recreation areas) and industrial. The multi-storey residential zone occupies floodplains on the right bank of the Modonkul River; the private housing zone is on its left bank and on the high floodplain of the right bank of the Dzhida River. The industrial zone includes: Pervomaysky and Inkursky mines, overburden rock dumps, the DTMMP areas, the Liteyshchik plant and the heat and power plant. All these facilities are located to the south of the town. Barun-Narynsky, Zun-Narynsky and Dzhidinsky tailing ponds are to the west of the residential zones on the right bank of the Modonkul River. The Modonkulsky mining field of technogenic sand was formed on its left bank. Top surfaces and gentle slopes of the hills in the Modonkul River valley outside the town limits were considered to be background territories.

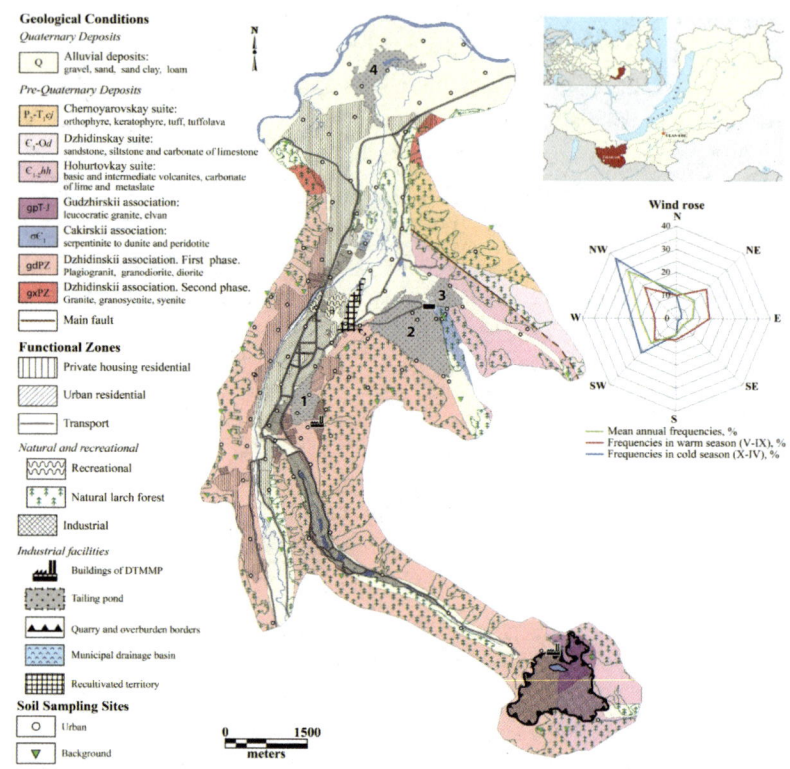

Fig. 1. Geological conditions of the town of Zakamensk and its functional zoning with the topsoil (0-10 cm) sampling sites. *Tailing ponds*: 1 - Dzhidinsky (piled), 2 - Barun-Narynsky (hydraulic-mine dump), 3 - Zun-Narynsky; 4 - the Modonkulsky mining field of technogenic sand.

Materials and methods

Soil-geochemical survey of the town of Zakamensk was carried out in summer 2012. All 129 samples (including 24 background ones) were taken from the topsoil (0-10 cm) on a regular grid with a step of 450-650 m (Fig. 1). Total concentrations of 54 elements were analyzed using inductively coupled plasma mass *spectrometry* (ICP-MS) at the Fedorovsky All-Russian Research Institute of Mineral Resources. Priority pollutants of hazard categories I (Zn, As, Pb, Cd), II (Cr, Co, Ni, Cu, Mo, Sb) and III (V, Sr, Ba, W) and some other elements (Be, Sn, Cs, Bi) were analyzed more in detail.

The values for HMs and metalloids in background samples C_b were grouped according to soil-forming rocks and compared to average global values in soils C (Kabata-Pendias, 2011) by calculating regional enrichment and depletion factors $REF= C_b/C$, $RDF=C/C_b$. Basing on the average values of REF/RDF geochemical spectra were drawn which characterize the geochemical features of background soils developed on different rocks.

Geochemical transformation of urban soils was also evaluated taking into account the lithologic-geochemical heterogeneity of the territory: local enrichment $LEF=C_i/C_b$ and depletion $LDF=C_b/C_i$ factors of elements (C_i is concentration of an ith element in urban samples) were calculated against the background values of soils formed on the same rocks. Geochemical spectra of urban soils are drawn for particular functional zones differing in both concentrations and combination of priority pollutants.

Environmental-geochemical assessment of urban soils is based on the indicators recommended for the complex environmental-geochemical and sanitary-hygienic investigation of urban environment, i.e. maximum and estimated permissible concentrations (MPC/EPC) (GN 2.1.7.2041-06, GN 2.1.7.2042-06) and the ratio showing the excess of maximum allowable concentrations of particular elements $I_r=C_i/MPC$.

The degree of the technogenically-induced geochemical transformation of soil cover was determined basing on the integral pollution index $Z_c=\sum LEF-(n-1)$, where n is the number of elements with $LEF>1$. The environmental hazard of high concentrations of HMs and metalloids in soils was assessed by the integral indicator accounting for the toxicity of pollutants: $IIR=\sum (p^*C_i/C)-(n-1)$, where n is the number of elements with $C_i > C$, p – the toxicity coefficient (for hazard category I elements p = 1,5; for hazard category II – 1,0; for hazard category III – 0,5) (Methodical recommendations ..., 1999). Values of Z_c and IIR are given in Table 1. Application of two integral indicators is reasonable because mining landscapes are located within a metallogenic anomaly with increased concentrations of mining and accompanying elements.

Z_c and IIR values	Level of contamination	Environmental risk
< 16	Low	Acceptable
16 – 32	Medium	Moderate
32 – 64	High	High
64 – 128	Very high	Very high
> 128	Maximum	Extremely high

Table 1. Levels of soil contamination with HMs and metalloids and corresponding degrees of environmental risk (Geochemistry…, 1990; Kasimov et al., 2012)

The natural and anthropogenic factors influencing spatial distribution of pollutants in soils and technogenic superficial objects (TSO) are identified using the regression trees method in the S-Plus software package (MathSoft, 1999). The method allows to determine levels of total soil pollution Z_c under various combinations of factors, and to estimate their importance.

Soil-geochemical data were visualized using the kriging method in MapInfo 11.5 and Surfer 11 packages. A fragment of the state geological map (Zinovyeva et al., 2011) and the plan of the town based on space imagery available in the MSU Geoportal system formed the basis for the geochemical maps. To prevent overestimation of pollution the sites with extremely high concentrations of HM and metalloids manifold above their averages in soils of the town were excluded from interpolation (Methodical recommendations …, 1999). Such sites identified by the three-sigma rule are shown on the map as point anomalies.

Results and discussion

Geochemical features of background soils. Background soils are formed on (1) alluvial deposits; (2) the Paleozoic Dzhidinsky complex, its first phase (plagiogranites, granodiorites and diorites); (3) the Middle Permian – Low Triassic Chernoyarovskaya suite; (4) the Cambrian complex (ore-bearing rock). The soils formed on these deposits are characterized by different concentrations of HMs and metalloids in the topsoils (Table 2).

Values	Element hazard category																	
	I category				II category						III category				n/determined			
	As	Cd	Pb	Zn	Co	Cr	Cu	Ni	Mo	Sb	Ba	Sr	V	W	Bi	Be	Cs	Sn
Alluvial sediments (n=6)*																		
average, mg/kg	7,7	0,3	22,8	98,8	14,8	69,3	35,5	40,2	3,6	2,2	506,7	290,0	126,5	7,1	0,5	1,8	16,5	2,9
min, mg/kg	4,3	0,28	20	78	11	55	28	30	1,9	0,66	440	210	89	4,5	0,38	1,4	3,6	1,9
max, mg/kg	16	0,38	29	120	18	88	48	55	5,4	5,6	550	390	150	9,9	0,66	2,3	42	3,6
REF/RDF**	1,1	1,3	1,2	1,4	1,3	1,2	1,1	1,4	3,2	3,2	1,1	1,7	1,0	4,2	1,3	1,4	3,3	1,2
Paleozoic Dzhidinsky complex (n=15)																		
average, mg/kg	4,9	0,4	31,7	133,5	22,8	47,8	81,0	34,1	7,8	1,6	698,0	274,0	175,5	34,3	1,3	3,8	10,8	2,6
min, mg/kg	3	0,23	18	34	11	37	34	20	1,4	0,71	380	130	81	3,1	0,22	1	4,3	1,4
max, mg/kg	6,7	0,72	100	200	38	67	200	67	48	3,1	1200	480	300	220	9,8	35	54	3,8
REF/RDF	1,4	1,1	1,2	1,9	2,0	1,2	2,1	1,2	7,1	2,4	1,5	1,6	1,4	20,2	3,0	2,8	2,1	1,1
Cambrian complex: Khokhurtovskaya and Dzhidinskaya suites (n=4)																		
average, mg/kg	3,5	0,5	40,5	113,0	13,6	35,4	39,3	27,8	5,1	1,2	587,5	205,0	101,0	31,2	1,1	1,2	5,3	1,7
min, mg/kg	1	0,23	5,9	37	3,3	8,6	27	11	2,1	0,64	290	150	14	2,1	0,19	0,39	1,1	1,3
max, mg/kg	5,4	0,89	90	200	18	62	67	48	10	2	1100	320	170	62	2,9	1,8	8,1	2,1
REF/RDF	2,0	1,1	1,5	1,6	1,2	1,7	1,0	1,0	4,6	1,8	1,3	1,2	1,3	18,3	2,7	1,1	1,0	1,4
Middle Permian – Low Triassic Chernoyarovskaya suite (n=2)																		
average, mg/kg	6,6	0,2	16,5	81,5	17,0	56,5	29,5	25,5	1,4	2,8	655,0	130,0	185,0	4,9	0,3	1,3	7,9	2,0
min, mg/kg	4	0,13	15	78	15	41	18	20	0,96	1,9	510	120	180	3,8	0,26	1	5,9	1,7
max, mg/kg	9,1	0,18	18	85	19	72	41	31	1,8	3,7	800	140	190	6	0,39	1,5	9,8	2,3
REF/RDF	1,0	2,6	1,6	1,2	1,5	1,1	1,3	1,1	1,3	4,2	1,4	1,3	1,4	2,9	1,3	1,1	1,6	1,3

Notes: *n – number of samples, **regional enrichment factors REF are in bold; regional depletion factors RDF are in italics.
Table 2. Statistical and geochemical parameters of HM and metalloid concentrations in the background topsoils (0-10 cm)

All background soils under study have a common group of elements with concentrations above the global percentage abundance, namely W (REF=2.9-20.2), Mo (1.3-7.1), Bi (2.7-3.7), Sb (1.8-4.2). Mining elements W and Mo are characterized by the highest values of REF. Soils on the Chernoyarovskaya suite formed of intermediate and basic effusive rocks are poor in all elements, except W, Mo, Sb, Cs, Co, Ba and V.

Soils on the Dzhidinsky complex of the Paleozoic age have the highest average concentrations of the elements of hazard categories I (Be and Zn); II (Co, Cu and Mo) and III (Ba and W). Background samples of soils on alluvial deposits are characterized by the greatest average concentrations of both hazardous elements (As, Cr, Ni, Sr) and those of non-determined hazard category (Bi, Cs, Sn). The lowest average concentrations of practically all elements are typical for soils on the Chernoyarovskaya suite and the Cambrian complex, except Sb and V for the first one and Cd and Pb for the second.

Geochemical features of urban soils. Priority pollutants of the topsoils in the town of Zakamensk are determined using the geochemical spectra for different functional zones. The elements are ranged according to the decrease of their

LEF values in the most polluted industrial zone (Fig. 2). Priority pollutants of soils of the industrial zone are as follows (the lower index is the *LEF* value): $Bi_{23.6}W_{21.0}Cd_{10.8}Be_{8.1}Pb_{8.0}Mo_{6.9}Sb_{6.6}$. The combination is due to the following sources:

> • tailing ponds with high concentrations of the above-listed elements inherited from the initial ores (Khodanovich et al., 2002a);
>
> • emissions of the Liteyshchik plant containing W, Sb, Mo, Pb, Cu, Cr (Geochemistry ..., 1990);
>
> • emissions of the heat and power plant which fire the fuel oil with mineral components obtaining V, Ni, Cr, Mo, Pb, Cu (Novoselov, 1983; Geochemistry ..., 1990).

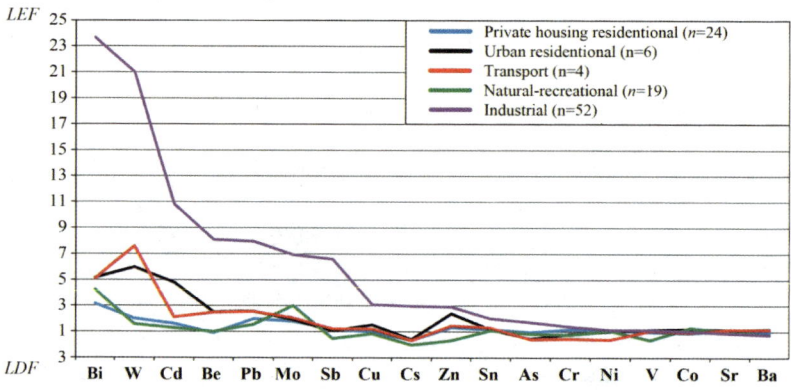

Fig. 2. Geochemical spectra of soils in different functional zones of the town of Zakamensk

The multi-storey residential zone is the second highest in soil contamination. Technogenic sands of the Dzhidinsky tailing pond lacking the surface vegetation cover are in the immediate vicinity of it. Therefore the intensive wind and water erosion, sheet washing and chemical (sulfate) weathering resulting in the accelerated oxidation of sulfides and dissolution of weathering products are observed there. The lateral migration leads to the enrichment of soils of the adjacent residential zone with $W_{6.0}Bi_{5.2}Cd_{4.8}Pb_{2.6}Be_{2.5}Zn_{2.4}$. Accumulation of $W_{7.6}Bi_{5.1}Be_{2.5}Pb_{2.5}Cd_{2.1}Mo_{2.1}$ is characteristic for the soils of the transport zone and $Bi_{4.3}Mo_{3.0}$ for those of the natural-recreational zone. The soils of private housing and natural-recreational zones are the least contaminated.

Hence, the chalcophylic elements (W, Bi, Cd, Pb and Mo), mainly in the form of sulfide minerals (Greenwood, Earnshaw, 2008), are among the priority pollutants of soils in the town of Zakamensk.

Assessment of the technogenic geochemical transformation of urban soils. Effects of the technogenic impact on soil cover of the town were evaluated against the background soils on the same soil-forming rocks using the map of the integral pollution index (Z_c) by HMs-metalloids complex.

The analysis of the map showed several anomalies (TSO or their combinations with natural soils) in the zone of the DTMMP influence. The most contrast one is within the emergency tailing pond reclaimed in 2011. The Z_c values in the exposed (formerly buried) humus horizon reach 485-721, 3-5 times above the extremely dangerous level.

The second anomaly was formed within the Barun-Narynsky and Zun-Narynsky tailing ponds with maximum Z_c values 316 and 292 respectively. The third one includes the Modonkulsky technogenic field on the left bank of the river; it was formed by the processes of washout, channel transportation and re-deposition of material from existing tailing ponds. The maximum Z_c values reach 200 there. The fourth anomaly covers the Dzhidinsky tailing pond and the emergency channel of DTMMP waste dumping where the Z_c values are more than twice above the extremely dangerous level.

The greatest technogenic influence on soils is characteristic of the areas where the DTMMP wastes are stored: the multi-storey residential zone with the highest densities of population, children's institutions and other infrastructure, floodplains and terraces of Modonkul, Barun-Naryn, Zun-Naryn. Inkur and other rivers, as well as the overburden rock dumps. The maximum level of soil and TSO contamination ($Z_c > 128$) is recorded within 26% of the urban territory where wastes containing toxic elements of hazard categories I to III are widely used (preparation of cement mix for construction of houses, road filling, etc.). These total TM concentrations result in the increased number of cases of respiratory and bone and muscular system diseases. Soils of private housing and the natural-recreational zone (suburban areas, both banks of the Modonkul River in its upper courses and the left bank in its middle and lower courses, upper courses of the dried-up Zun-Naryn and Barun-Naryn rivers) show the lowest technogenic geochemical transformation ($Z_c > 16$). The area of these sites accounts for 35% of the whole territory.

Assessment of the environmental risk associated with the soil cover pollution in Zakamensk. To assess the environmental hazard of soil contamination in the town of Zakamensk the concentrations of TM and metalloids were weighted against the RF standards for V, Pb, As, Cd, Cu, Ni, Zn and Sb.

Pb, Sb, Cd and As are among the most hazardous elements in all functional zones (Fig. 3). Their distribution in soil cover depends mainly on the level of technogenic loading and the distance from tailing ponds. Thus, in the industrial zone the I_r coefficients for these elements are 7.8, 3.6, 2.1 and 1.7 respectively, while in the multi-storey residential one 1.8, 0.5, 0.8 and 1.0. The lowest I_r values for all elements, except Pb, Zn and Sb, are characteristic of the soils of the transport zone. The main sources of pollution are DTMMP wastes because the W-Mo ores of the Pervomaysky stockwork, Inkursky and Holtosonsky fields are also rich in Pb (400 to 5600 mg/kg) and Zn (380 to 3800 mg/kg) (Smirnova, Plusnin, 2013).

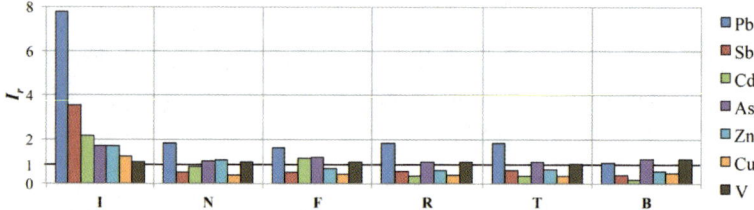

Fig. 3. I_r ratio showing the excess of maximum permissible concentrations of heavy metals and metalloids in soils of the functional zones of the town of Zakamensk. Zones: I – industrial; N – multi-storey residential; F – private housing; R – natural-recreational; T – transport. B – background territories.

Multi-element pollution of topsoils causes very dangerous environmental situation in the town of Zakamensk: values of the *IIR* integrated index vary from 1.5 to 1737.5 with an average of 93.3. There are several anomalies within the town (tailing ponds, the Modonkulsky technogenic field, and the dammed valley of the Inkur River). Nearly half of the town (49%) is within the zone of environmental disaster with recorded changes in morbidity rates for children and teenagers (Prusakov et al., 2005).

Thus, a critical environmental situation with very dangerous and extremely dangerous contamination of soils is characteristic of two-thirds of the urban territory. Reclamation of Dzhidinsky, Barun-Narynsky, Zun-Narynsky tailing ponds and the Modonkulsky field with isolation or phyto-remediation of DTMMP wastes could be considered among the urgent measures for improvement of the environmental situation. The environmental monitoring of the new Zun-Narynsky tailing pond and adjacent territories is necessary to control the dynamics of accumulation of HMs and metalloids.

Conclusions

1. Priority pollutants of soils in the town of Zakamensk are chalcophylic elements W, Bi, CD, Pb and Mo. This is caused by the active processes of physical and chemical aeration of tailing ponds, and also by the use of wastes from the Kholtosonsky and Inkursky sulphide-tungsten ore fields for construction of buildings, playgrounds, roads, etc.

2. Spatial distribution of the integral pollution index (Z_c) depends on the degree of technogenic impact, specific features of extraction technologies, as well as landscape and geochemical factors. The maximum Z_c values (185-710) are characteristic of the TSO with low content of clay fraction and acid reaction (Barun-Narynsky, Zun-Narynsky, Dzhidinsky and emergency tailing ponds, the channel of waste dumping, the Modonkulsky technogenic field). In other functional zones HMs accumulate in the subordinated positions and on the biogeochemical barrier which capacity depends on the amount of humus. These sites with average Z_c of 36.7 are located within the floodplain and the first terrace of the Modonkul River.

3. As a result of the DTMMP activities the critical environmental situation is characteristic of two-thirds of the urban soil cover. Pb, Sb, Cd and As which concentrations in soils of the industrial zone exceed maximum and estimated permissible concentrations (MPC/EPC) by 1.7-7.8 times are among the most hazardous elements. Reclamation measures with isolation or phyto-remediation of DTMMP wastes are necessary, as well as the geochemical monitoring of soils and vegetation in the impact zone of tailing ponds and ore fields.

4. Experience of the environmental-geochemical assessment of soil cover in the impact zone of mining enterprises could be useful for other fields of the non-ferrous metals with high lithological-geochemical heterogeneity of the ter-

ritory. It suggests the need of accounting for the geological diversity and specific features of metallogeny of an area. Geochemical indices *LEF/LDF* should be calculated against the individual background values for each soil-forming rock. Such approach allows more accurate assessment of the degree of technogenic geochemical transformation of soils and the environmental hazard of pollution.

References

Авессаломова И.А. (1986) Ландшафтно-функциональные карты при изучении геохимических аномалий в городе // Вестник Моск. ун-та. Сер. География. № 5. С. 88-94. Publication in Russian language [Avessalomova I.A. (1986) Landscape-functional maps for studying geochemical anomalies in a city// Bulletin of the Moscow Univ. Ser. Geography. No. 5. P. 88-94.]

Авессаломова И.А. (2004) Функционирование и динамика горных ландшафтов / География, общество, окружающая среда. Том II. Функционирование и современное состояние ландшафтов. М.: Городец. С. 154-170. Publication in Russian language. [Avessalomova I.A. (2004) Functioning and dynamics of mountain landscapes / Geography, Society, Environment. Volume II. Functioning and Current State of Landscapes. M.: Gorodets. P. 154-170.]

Геохимия окружающей среды (1990)/ Ю.Е. Сает, Б.А. Ревич, Е.П. Янин и др. М.: Недра, 335 с. Publication in Russian language. [Geochemistry of Environment (1990)/ Yu.E. Sayet, B.A. Revich, E.P. Yanin, et al. M.: Nedra, 335 pp.]

Гринвуд Н.Н., Эрншо А. (2008) Химия элементов. В 2-х томах. М.: Изд-во «БИНОМ». 1267 с. Translation into Russian [Greenwood N. N., Earnshaw A. (2008) Chemistry of the elements. In 2 volumes. M.: Publishing House "BINOMIAL". 1267 pp.]

Добровольский В.В., Урусевская И.С. (2004) География почв. М.: Изд-во МГУ, Изд-во "КолосС". 460 с. Publication in Russian language. [Dobrovolsky V. V., Urusevskaya I.S. (2004) Geography of soils. M.: MSU Publishing House, Publishing House "Colossus". 460 pp.]

Касимов Н.С., Битюкова В.Р., Кислов А.В. и др. (2012) Проблемы экогеохимии крупных городов // Охрана и разведка недр, № 7, с. 8-13. Publication in Russian language. [Kasimov N.S., Bityukova V.R., Kislov A.V., et al. (2012) Ecogeochemistry Issues of Large Cities//Prospects and protection of mineral resources, No. 7, p. 8-13.]

Касимов Н.С. (2013) Экогеохимия ландшафтов. М.: ИП Филимонов М.В. 208 с. Publication in Russian language. [Kasimov N. S. (2013) Landscape Ecogeochemistry. M.: IP Filimonov M. V. 208 pp.]

Методические рекомендации по оценке загрязненности городских почв и снежного покрова тяжелыми металлами (1999)/ Сост. В.А. Большаков, Ю.Н. Водяницкий, Т.И. Борисочкина и др. М.: Почвенный институт им. В.В. Докучаева. 32 с. Publication in Russian language. [Methodical recommendations on the assessment of urban soils and snow cover pollution with heavy metals (1999)/ Compiled by V.A. Bolshakov, Yu.N. Vodyanitsky, T.I. Borisochkina, et al. M.: V.V.Dokuchayev Soil Institute. 32 pp.]

Ногина Н.А. (1964) Почвы Забайкалья. М.: Наука, 312 с. Publication in Russian language. [Nogina N. A. (1964) Soils of Transbaikalia. M.: Nauka, 312 pp.]

Новоселов С.С. (1983) Исследование выбросов в атмосферу твердых продуктов сгорания мазута и разработка методов их сокращения. Дисс. ... к.т.н. М., 171 с. Publication in Russian language. [Novoselov S. S. (1983) Investigation of emissions of solid products of fuel oil combustion into the atmosphere and development of methods of their reduction. Thes. ... PhD. in Tech.Sci. M., 171pp.]

Прусаков В.М., Прусакова А.В., Басараба И.Н. и др. (2005) Оценка риска здоровью детского населения от воздействия техногенных песков вольфрамово-молибденового ГОКа // Методологические проблемы экологически обусловленных нарушений здоровья. Бюллетень ВСНЦ СО РАМН. Иркутск: НЦРВХ СО РАМН, с. 55-60. Publication in Russian language. [Prusakov V.M., Prusakova A.V., Basaraba I.N., et al. (2005) Assessment of the risk to health of the children's population from the exposure to technogenic sands of a tungsten-molybdenum mining plant//Methodological problems of the environmentally-caused health deterioration. Bulletin of the VSNTs, SB of the Russian Academy of Medical Science. Irkutsk: NTsRVKh, SB of the Russian Academy of Medical Science, p. 55-60.]

Смирнова О.К, Плюснин А.М. (2013) Джидинский рудный район (проблемы состояния окружающей среды). Улан-Удэ: Издательство Бурятского научного центра СО РАН. 181 с. Publication in Russian language. [Smirnova O. K., Plusnin A.M. (2013) The Dzhidinsky ore district (environmental issues). Ulan-Ude: Publishing House of the Buryat scientific center of the Siberian Branch of the Russian Academy of Science. 181 pp.]

Удачин В.Н. (2012) Экогеохимия горнопромышленного техногенеза Южного Урала. Дисс... д.г.-м.н. Миасс, 352 с. Publication in Russian language. [Udachin V. N. (2012) Ekogeochemistry of mining technogenesis in the Southern Urals. Thes... D.Sc. in Geology. Miass, 352 pp.]

Уфимцев Г. Ф. (1991) Горные пояса континентов и симметрия рельефа Земли. Новосибирск: Наука. Сиб. отд-ние, 168 с. Publication in Russian language. [Ufimtsev G. F.(1991) Mountain belts of the continents and the symmetry of the Earth's relief. Novosibirsk: Nauka. Sib. Br., 168 pp.]

Ходанович П.Ю., Смирнова О.К., Яценко Р.И. (2002) Экологические проблемы освоения сульфидсодержащих вольфрамовых месторождений в условиях таежно-мерзлотных ландшафтов расчлененного среднегорья // Горный информационно-аналитический бюллетень. № 12, с. 52-59. Publication in Russian language. [Khodanovich P.Yu.,

Smirnova O.K., Yatsenko R.I. (2002) Environmental problems of the development of sulfide-containing tungsten deposits within taiga permafrost landscapes of dissected middle mountains//the Mountain information and analytical bulletin. No. 12, pp. 52-59.]

Kabata-Pendias A. (2011) Trace Elements in Soils and Plants. Fourth Edition. CRC Press, 548 p.

Makhinova A.F., Makhinov A.N., Kuptsova V.A., Liu Shuguang, Ermoshin V.V. (2014) Landscape–Geochemical Zoning of the Amur Basin (Russian Territory) // Russian Journal of Pacific Geology, vol. 8, №2, pp. 138-150.

Нормативные акты:

ГН 2.1.7.2041-06 «Предельно допустимые концентрации (ПДК) химических веществ в почве», утвержден 19 января 2006 г. Publication in Russian language. [GN 2.1.7.2041-06 "The Maximum Permissible Concentrations (MPC) of Chemicals in Soil", approved on January 19, 2006.]

ГН 2.1.7.2042-06 «Ориентировочно-допустимые концентрации (ОДК) химических веществ в почве», утвержден 9 января 2006 г. Publication in Russian language. [GN 2.1.7.2042-06 "The Estimated Permissible Concentrations (EPC) of Chemicals in Soil", approved on January 9, 2006.]

Environmental-Geochemical Map of Ulaanbaatar City: Methodology of Compiling and Perspectives of Applying

Olga Sorokina[1]

[1] Faculty of Geography, Lomonosov Moscow State University, Moscow, Russia

Abstract

The wide spectrum of spatial and analytical data was used to compile an integral map which combined environmental, geochemical and anthropogenic information about the urban territory. The 20 types of urban landscapes were defined within Ulaanbaatar city. They differed from original natural areas to significantly transformed industrial territories. The analysis of the location of pollution fields in geomedias showed the areas with a long-term contamination and with year-round and seasonal air pollution. Moreover the map allowed us to trace the migration of pollutants between geomedia, to forecast future changes of the pattern of urban and suburban territories, and to assess the potential danger of geomedia pollution in Ulaanbaatar.

1. Introduction

The purpose of the study is to compile an environmental-geochemical map of urban territory which includes both landscape and technogenic information and shows the result of anthropogenic impact.

The object of the study is Ulaanbaatar city. It's a rapidly growing capital of Mongolia with population about 1.3 mln. The main anthropogenic sources are three thermal power plants (TPPs) using brown coal with a high content of heavy metals (Kosheleva et al. 2010), industrial and municipal waste, and 260 thousand vehicles using leaded petrol (Sorokina, Enkh-Amgalan 2012). A regional feature of Ulaanbaatar is ger (or yurt) quarters – residential areas with traditional Mongolian dwellings and private houses where half of the city population lives and uses brown coal for heating (Gunin et al. 2003; Gutticunda 2008).

The city is located in a wide intermountain valley of Tuul river, within a unique forest-steppe ecotone natural zone (Gunin et al. 2003; Ecosystems 2005). This area has a severely continental climate with winter anticyclonic weather conditions which cause temperature inversions and increase the air pollution (Kasimov et al. 1995).

2. Data Sources and Mapping Procedure

The map is based on a field work and chemical-analytical data obtained in 2008-2012 by the author and colleagues from Faculty of Geography of Lomonosov MSU and Joint Russian-Mongolian Complex Biological Expedition of RAS and MAS.

Landscapes and land use of Ulaanbaatar city and the surrounding area (up to 50 km from the city) were studied using satellite images (URL 1; URL 2), maps (e.g. National... 1990) and published materials (e.g. Nogina 1984; Batkhishig 1999; Gunin 2003; Ecosystems... 2005). The collected data were verified and refined during field work.

Geomedia sampling included snow cover (21 samples), topsoils (96) and poplar leaves (84) of urban and background areas. In all the samples, concentrations of heavy metals were determined by ICP-MS method. The detailed information about sampling process, samples location, geochemical features and pollution of Ulaanbatar city geomedia were published in (Kasimov et al. 2011a, 2011b; Sorokina et al. 2013).

Present-day methods of mapping of the urban environment were studied (e.g. Mapping... 2011; Kasimov et al. 2013). ArcGIS 10.0 was used for mapping in this study.

The map compiling algorithm is shown on the fig. 1. It involves three steps:
- a map of the modern city territory structure compiled using spatial data from field observations, satellite images, maps, literature etc.;

- a map of pollution sources and pollution fields in geomedia compiled using analytical and published data;

- a complex map with landscape, land use, geochemical, and pollution information.

3. Results and discussion

The main result of the work is the complex environmental-geochemical map of Ulaanbaatar city and surrounding areas (fig. 2). The legend of this map is based on the principles of geochemical systematics of urban elementary landscapes. It consists of information about the modern structure of the territory (matrix I), the assessment of the anthropogenic impact (matrix II) and the list of main pollution sources (matrix III).

3.1. The pattern of the modern city territory

The 20 types of urban landscapes form the pattern of the modern Ulaanbaatar city. They differ with the original natural features (fig. 2, columns of matrix I) and the degree of anthropogenic transformation (fig. 2, lines of matrix I).

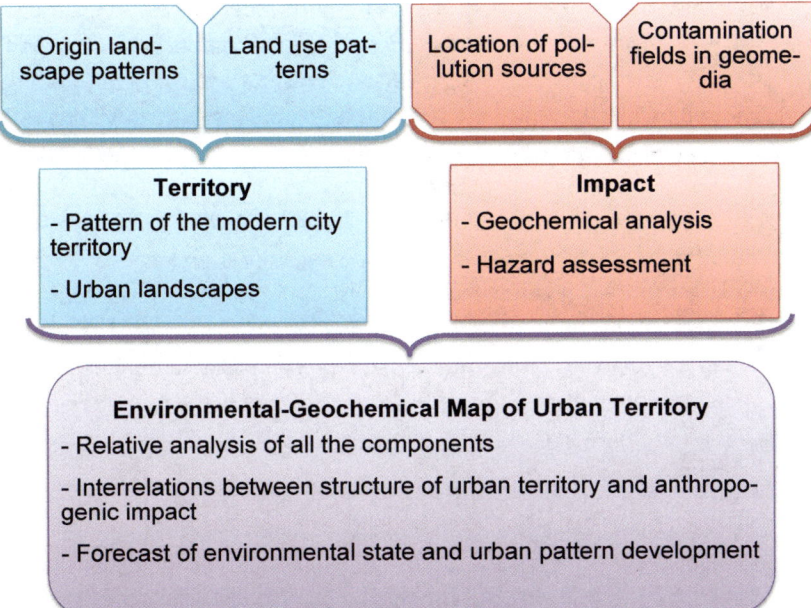

Fig.1. Environmental-geochemical mapping algorithm.

Taiga, steppe and flood plain landscapes can be distinguished quite clearly even in the modern city. They have been modified in accordance with human activity. In the Tuul flood plain, an industrial-traffic-residential urban center with the multistoried residential areas, plants and factories, main squares and roads, and with a high degree of soil sealing has formed. On the steppe slopes, vast ger quarters with private residential building and with a low degree of soil sealing are located. On the taiga slopes, there is a recreation area of Bogd Khaan Uul Mountain which is one of the oldest reserves in the world.

The degree of impact and transformation of the original natural landscape increases in the range: recreation → ger → multistoried → industrial area. Urban landscapes in recreational series (fig. 2, line R in matrix I) largely correspond to the initial natural analogues. Anthropogenic influence can be expressed in the compaction of the soil surface, the reduction of the vegetation, the invasion of weed species, and in garbage dumps.

Urban landscapes of ger series (line G) are characterized by dense private residential buildings (houses and gers), the almost complete destruction of woody vegetation, and the replacing for the original herbaceous species to the weed and resistant to trampling species (*Carex duriuscula, Atriplex patula*, etc.). Soil compaction (paths, ger sites), mixing of the upper soil horizons (e.g. vegetable gardens), the extensive network of dirt roads, and single paved roads are also observed in ger quarters.

Urban landscapes of multistoried (line M) and industrial series (line I) are transformed significantly compared to the original natural landscapes. Technological soils are formed instead of natural ones, the sealed soils area increases, the radial and lateral water migration ratio changes. The woody vegetation is represented by planted poplar (*Populus laurifolia*) and larch trees (*Larix sibirica*). In the industrial area the impact is higher than in the multistoried one, due to the lower proportion of green space and a greater variety and intensity of pollution sources.

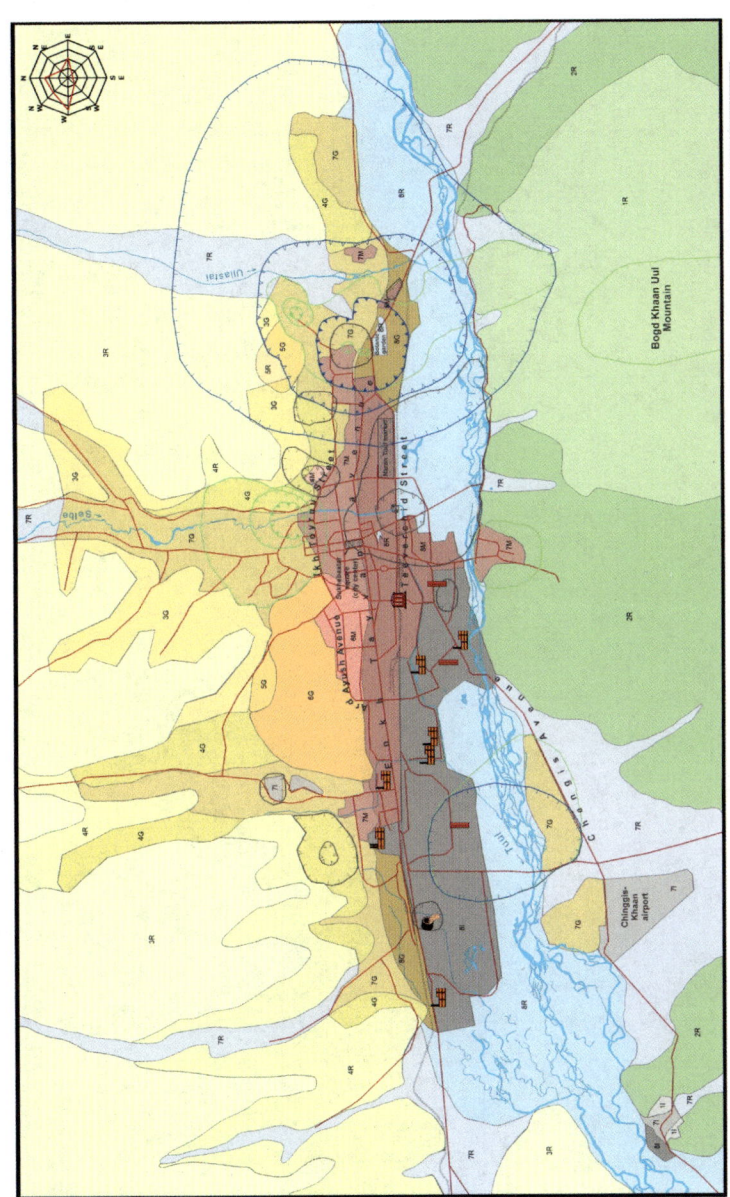

Fig.2. Environmental-Geochemical Map of Ulaanbaatar City

I. Landscapes and Land Use of Ulaanbaatar City

		Siberian pine-larch dwarfshrub (marsh labrador tea, lingonberry) greenmoss mountain boreal (taiga) forests	Larch herb-sedge and herb-grass mountain hemiboreal forests	Herb-sedge-bluegrass-meadow and herb-bunchgrass steppe		Shrubby (mostly poplar-willow) herb-grass with halophyte-grass and sedge species flood-plains	
Vegetation							
Soils		Mountain soddy forest, mountain taiga cryogenic	Mountain soddy forest, mountain meadow-forest	Mountain chestnut and dark chestnut, mountain chernozems		Alluvials stony-pebble	
Parent rocks		Archean granites	Carboniferous shales	Carboniferous shales	Neogen clay	Quaternary river alluvium	
Relief		Nothern slopes of intermountain valley		Southern slopes of intermountain valley		Flood plains of Tuul river tributaries	Flood plains of Tuul river
Landscape-geochemical position in catena		Transeluvial		Transeluvial	Trans-accumulative	Transeluvial	Trans-accumulative	Transaccumulative-superaqual	
		1	2	3	4	5	6	7	8
Land use	Recreation, R								
	Residential, ger quarters, G								
	Residential, multistoried, M								
	Industrial, I								

II. Environmental-Geochemical Assessment of Ulaanbaatar Geomedia

Geomedia	Snow and soils pollution levels. Biogeochemical transformation of plants					
	Low / Not dangerous	Medium / Low dangerous	High / Dangerous	Very high / Very dangerous		
Solid fraction of snow water (snow dust), Z_c	< 32	32 - 64	64 - 128	> 128		
Topsoils, Z_c	< 16	16 - 32	32 - 64	> 64		
Poplar leaves (*Populus laurifolia*), Z_v	< 15	15 - 20	20 - 25	> 25		

Note
Z_c is an Integral Soil / Snow Pollution Index: $Z_c = \Sigma\ (EF > 1.5) - (n - 1)$, where EF – Enrichment factors of heavy metals, n - amount of heavy metals with $EF > 1.5$

Z_v is an Integral Biogeochemical Index: $Z_v = \Sigma\ (EF > 1.5) + \Sigma\ (1/EF > 1.5) - (n - 1)$, where EF - Enrichment and $1/EF$ - biodisperation factors of heavy metals, n - amount of heavy metals with $EF > 1.5$ and $1/EF > 1.5$

III. Pollution Sources

Industrial plants and factories

Thermal power plants

Wastewater treatment plant

Main railway station

Railways

Roads

Other

Rivers

Main squares and markets

The industrial-traffic-residential frame of Ulaanbaatar city had formed during last few decades. Contemporary building activity is adjusting to it. The main processes of the modern city development are:

- an expansion of ger quarters to the north of the city where areas 3R are being transformed to the 3G, areas 4R to the 4G, areas 7R to the 7G;

- an active multistoried building process to the south and southeast of the city center where areas 8R are being transformed to the 8M.

- In prospect this expansion will continue and also territories to the south and southwest of the industrial area (8R, 7G) will be changed.

3.2. The assessment of the anthropogenic impact

Based on the average contents of heavy metals in the solid and liquid fractions of snow water (snow dust and snow water respectively), topsoils, and poplar leaves in Ulaanbaatar city and surrounding territory we calculated enrichment and dispersion factors (*EF* and *1/EF* respectively) and integral indexes (*Zc* and *Zv*) for every sampling point and every geomedia (fig. 2, matrix II). The detailed geochemical and hazard analysis of geomedia were published in (Kasimov et al. 2011a, 2011b; Sorokina et al. 2013). For the complex map compiling we were interested in the integral indexes only (fig. 2, matrix II). Hazard criteria are taken from (Environmental 1990; Kasimov et al. 2011b).

The snow contamination reflects the air pollution during winter when the impact on the Ulaanbaatar landscapes is the strongest due to the heating season. The city average Integral Pollution Index *Zc* for the snow dust is 30, which corresponds to the low level of contamination. At the same time *Zc* for the snow water is 559, which means the highest pollution level. The snow water is strongly polluted with the dissolved forms of heavy metals almost on the whole city territory. It is a potential risk of contamination of groundwater and surface water with mobile forms of heavy metals. By the end of the winter medium- and high-contrast geochemical anomalies of heavy metals formed in the center and east of the city, and in the western part of the industrial area. The most significant impact was found in ger quarters (*Zc* = 853 for the snow water and 41 for the snow dust) and industrial areas (547 and 42 respectively).

The contamination of poplar leaves reflects the features of the air transport of pollutants in summer. The city average Integral Biogeochemical Index Zv is 13 and various slightly with the trace elements imbalance of 10 - 15 units in most parts of the city. The strongest impact with abnormal for Ulaanbaatar value of Zv = 49 was found along the roads, the weakest one with Zv = 5 was found in the recreation area. On the map there are three types of biogeochemical anomalies: 1) caused by traffic – in the north of the city center, 2) caused by industry – in the industrial area, near TPPs, 3) caused by the intensive air transport of pollutants – in the east of the city, the Uliastai area. On the Sukhbaatar square a minimum biogeochemical transformation is observed because of a young tree planting and beautification activity in a tourist part of the city.

The topsoil contamination characterizes long-term trends of the city pollution. The Ulaanbaatar average Integral Pollution Index Zc for topsoils is 11, which corresponds to the low level of contamination. Zc is 16 – 32 units for the topsoils of the industrial area, city center and near busy roads. A western part of the city is weakly polluted $(Zc < 8)$. In ger quarters geochemical anomalies reach 30 units due to the traffic and household waste dumps. Near the sewage treatment plant Zc is 26.

In Ulaanbaatar city all the main sources are located in areas 8I, 3-8G and along the roads. But the strongest geochemical anomalies were found in areas 7G, 7M, 8G in the east of the city. The main reason is that western wind influences significantly on the air migration of pollutants within Ulaanbaatar city. So the location of anthropogenic geochemical anomalies often does not coincide with the location of pollution sources.

Despite the fact that Bogd Khaan Uul Mountain is a protected area, pollution with heavy metals is found on its northern slope. This fact was also reported by (Kasimov 1995). The reasons are the strong urban air contamination and air circulation features within the intermountain valley.

3.3. The relative analysis of contamination fields in geomedia

The relative analysis of contamination fields in geomedia reviles pollution trends. Matching geochemical anomalies in the snow cover and poplar leaves indicate areas of a year-round immission whereas non-matching anomalies indicate summer (leaves) or winter ones (snow). Contamination of topsoils indicates areas of a long-term pollution.

The biggest long-term year-round geochemical anomalies forms in Ulaanbaatar in the confluence of the Uliastai and the Tuul (areas 7-8 R, G, M). A big year-round anomaly forms in the snow and leaves to the south of industrial area (7G, 8R, 8I). A vast anomaly of summer period forms in the lower reaches of the Selbe (4G, 7G) and a small one forms on the left bank of the Tuul (7M).

Large winter anomalies are observed in the snow cover of ger quarters on the northern slopes of the valley (3-4G, 7G). These anomalies are contaminated predominantly with the dissolved mobile forms of heavy metals which migrate with meltwater and accumulate in flood plain transaccumulative - superaqual geochemical positions of relief where wide long-term topsoil anomalies forms (areas 8R, M). The other long-term topsoil anomaly is found near wastewater treatment plant (8I) and caused by the sewage.

4. Conclusions

4.1. The environmental-geochemical map shows the modern pattern of the city territory. It allows to understand a present-day structure of the city and to estimate the degree of anthropogenic transformations. It also can be used to predict its future changes of the territory and to plan its development.

4.2. The location of anomalies of pollutants often does not coincide with the location of pollution sources. The map allows to define them, to analyze, and to compare different natural and anthropogenic migration factors.

4.3 The relative analysis of contamination fields in geomedia reviles pollution time trends. Geochemical anomalies in the snow cover and poplar leaves identify areas of year-round and seasonal heavy metals immission from the air. Topsoil contamination indicates areas of the long-term pollution.

4.4. The addition of new data about soil cover patterns and topsoil chemical characteristics (pH, the amount of the organic carbon, etc.) can be used to improve the map. This data can help to analyze geochemical migration features and to indicate areas that are potentially vulnerable for contamination.

Acknowledgements

Author would like to thank Faculty of Geography of Lomonosov Moscow State University and N.S. Kasimov, N.E. Kosheleva, E.M. Nikiforova, D.L. Golovanov, M.D. Bogdanova for their assistance at various stages of data processing and analyzing, Joint Russian-Mongolian Complex Biological Expedition of RAS and MAS and P.D. Gunin, S.N. Bazha, A.V. Andreev, S. Hadbaatar for the aid with field work, data collection and for the financial support, Institute of Geography of MAS and S. Enkh-Amgalan for the contribution to the field work and data collection.

References

Gunin P.D., Evdokimova A.K., Bazha S.N., and Sandar' M. (2003). Social and Ecological Problems of Mongolian Ethnic Community in Urbanized Territories. Moscow: Rossel'khozakademiya. P.1–59.

Gutticunda S. (2007). Urban air pollution analysis for Ulaanbaatar. The World Bank Consultant Report. Washington DC, USA. 125 p.

Kosheleva N.E., Kasimov N.S., Dorjgotov D., Baja S.N., Golovanov D.L., Sorokina O.I., Enkh-Amgalan S. (2010). Assessment of heavy metal pollution of soils in industrial cities of Mongolia. Geography, Environment, Sustainability. 3 (2): 51-65.

Kasimov N.S., Kosheleva N.E., Sorokina O.I., Bazha S.N., Gunin P.D., and Enkh-Amgalan S. (2011a). Environmental-geochemical state of soils in Ulaanbaatar (Mongolia). Eurasian Soil Science. 44 (7): 709-721.

Kasimov N.S., Kosheleva N.E., Sorokina O.I., Bazha S.N., Gunin P.D., and Enkh-Amgalan S. (2011b). An ecological-geochemical assessment of the state of woody vegetation in Ulaanbaatar city (Mongolia). Arid Ecosystems. 1 (4): 201-213.

Mapping the chemical environment of urban areas (2011). Ed. by Johnson C.C., Demetriades A., Locutura J., Ottersen R.T. UK: John Wiley & Sons, Ltd. 640 p.

Sorokina O.I., Enkh-Amgalan S. (2012). Lead in the landscapes of Ulaanbaatar city. Arid ecosystems. 2 (1): 61-67.

Sorokina O.I., Kosheleva N.E., Kasimov N.S., Golovanov D.L., Bazha S.N., Dorjgotov D., Enkh-Amgalan S. (2013). Heavy metals in the air and in the snow cover of Ulaanbaatar city. Geography and Natural Resources. 34 (3): 291-301.

Касимов Н.С., Лычагин М.Ю., Евдокимова А.К., Голованов Д.Л., Пиковский Ю.И. (1995). Улан-Батор, Монголия (теплоэнергетика). Межгорная котловина / Экогеохимия городских ландшафтов / Под ред. Касимова Н.С. М.: изд-во МГУ: 231-248. (In Russian). [Kasimov N.S., Lychagin M.Yu., Evdokimova A.K., Golovanov D.L. , Pikovsky Yu.I. (1995). Ulaanbaatar, Mongolia (heat power engineering). Intermountain depression. In: Ecogeochemistry of urban landscapes. Ed. by Kasimov N.S. Moscow: MSU Publishing. P. 231-248].

Батхишиг О. (1999). Почвенно-геохимические особенности долины р. Туул. Автореф. дисс. ... канд. геогр. наук. Улаанбаатар: Ин-т геоэкологии АН Монголии. 23 с. (In Russian). [Batkhishig O. (1999). Soil geochemical characteristics of the Tuul river valley. Candidate dissertation abstract. Ulaanbaatar: Institute of Geoecology MAS. 23 p.]

Геохимия окружающей среды. (1990). М.: Недра. 335 с. (In Russian). [Environmental geochemistry. (1990). Moscow: Nedra. 335 p.]

Касимов Н.С., Никифорова Е.М., Кошелева Н.Е., Хайбрахманов Т.С. (2013). Геоинформационное ландшафтно-геохимическое картографирование городских территорий (на примере ВАО Москвы. 2. Ландшафтно-геохимическая карта. Геоинформатика. № 1. С. 28-32. (In Russian). [Kasimov N.S., Nikiforova E.M., Kosheleva N.E., Khaibrakhmanov T.S. (2013). GIS landscape-geochemical mapping of urban areas (Eastern Administrative District of Moscow). Part 2. Landscape-geochemical map. Geoinformatica. no. 1. P. 28-32]

Национальный атлас Монгольской Народной Республики. (1990). Улан-Батор, М.: ГУГК ГСК МНР – ГУГК СССР. 144 с. (In Russian). [National atlas of Mongolia. (1990). Ulaanbaatar, Moscow: GUGK GSK of Mongolia – GUGK USSR. 144 p.]

Ногина Н.А. (1984). Почвенный покров и почвы Монголии. М.: Наука. 192 с. (In Russian). [Nogina N.A. (1984). Soils of Mongolia. Moscow: Nauka. 192 p.]

Экосистемы бассейна Селенги. (2005). Труды совместной Российско-Монгольской комплексной биологической экспедиции / отв. ред. Востокова Е.А., Гунин П.Д. М.: Наука. т. 44. 395 с. (In Russian). [Ecosystems of the Selenga river basin. (2005). In: Proceedings of the Joined Russian-Mongolian Complex Biological Expedition. Ed. by Vostokova E.A., Gunin P.D. Moscow: Nauka. Vol. 44. 395 p.]

URL 1: http://glcfapp.glcf.umd.edu (web resource of satellite images and satellite data)

URL 2: http://www.cgiar-csi.org/ (web site of Consortium for Spatial Information)

III. Fluvial transport dynamics and morphology

Source to Sink: Water and Sediment Transport in the Selenga-Baikal Catchment

Ekaterina Promakhova[1], Nikolay Alexeevsky[2]

[1] Faculty of Geography, Lomonosov Moscow State University, Moscow, Russian Federation, promakhova@gmail.com
[2] Faculty of Geography, Lomonosov Moscow State University, Moscow, Russian Federation, n_alex50@mail.ru

Abstract

The Selenga River is the main tributary of Lake Baikal. We analyzed the average annual water balance and related the average annual and seasonal sediment load budgets of the river and its tributaries. Determined that the average annual water flow at the mouth for the period from 1934 to 2009 was 28.6 km³. The water flow in the Selenga River at the Russian-Mongolian border was 37% of this value. The 73% and 7% of the average annual sediment load passed across the border during high water and low water stages, respectively. It was found that longitudinal suspended sediment concentration depended on the water regime phase, it decreased in 160 times for floods and increased in 15 times for low water.

Keywords – sediment load budget, water budget, suspended sediment concentration.

1. Introduction

The Selenga River is the major tributary of Lake Baikal, its length is 1024 km. The river provides 60% of the annual water inflow of the lake (Гармаев, Христофоров, 2010). The biggest tributary of the Selenga River is the Orkhon River that exceeds its length of 100 km. The total length of the river system from the Orkhon River source to the Selenga River mouth is 1546 km. The area of the Selenga River Basin is 447,000 km², the upstream part of the basin is located in Mongolia (67%), the downstream part belongs to Russian Federation (33%). There are spring snowmelt floods and summer rain floods for most of part of the basin. In the south-eastern rivers of the basin spring floods virtually lack, in the high mountain area snowmelt floods observed during spring and summer.

Issue of water resources and river water quality of the Selenga Basin is becoming urgent nowadays (Nadmitov et al., 2015; Pfeiffer et al., 2015; Thorslund et al., 2012). Low-flow period is marked in the study region since the 2000s (Потемкина, 2011). Deterioration has occurred in the winter-autumn low water stage in 2014-2015, thereby the Russian Government has decided to draw off the Baikal level below acceptable marks by the Irkutsk HPS (Decree No. 97 of 04.02.2015) with the subsequent impoundment of the lake in the spring flood 2015. Moreover, new approaches in Mongolian water management was set after the election of the new President in 2013 and Ministry of Nature reorganizing (Гомбоев et al., 2014). As a result, Mongolia has become once again discuss the Orkhon-Gobi water diversion project and construction of hydropower stations on the rivers of the basin (Withanachchi et al., 2014). To do this, the Mongolian side is attracting investments of the World Bank (Project No. P118109) (MN-Mining Infrastructure ..., 2013). The project may lead to irreparable damages for the ecosystem of the region, therefore Russia, in turn, wrote a letter to the bank to refrain from financing (River without Boundaries, 2013).

Development of the mining industry in the Selenga River Basin leads to decrease in water resources, water quality deterioration and additional fluxes of suspended solids and associated pollutants (Chalov et al., 2015; Karthe et al., 2014).

The main objective of this paper is analysis of the water and suspended sediment load budgets of the Selenga River Basin under the growing human impact in a sparsely monitored region.

2. Methods

Water flow data was obtained from the 18 gauging stations in Mongolia and 58 in Russia. The final site (Raz'ezd Mostovoy) is located at the top of the Selenga Delta in 115 km from the Baikal Lake, it was opened in 1934. The Novoselenginsk site found in 1932, it is the station with the longest observation period. Within the Russian part of the Selenga River Basin daily suspended sediment concentration (SSC) is determined by 12 gauges at the 6 rivers (Selenga, Chikoy, Khilok, Uda, Kiran, Kuytunka). In Mongolia occasional observations were conducted at the six sites (Tuul, Orkhon, Selenga, Khara, Eroo, Egeyn Gol) (Stubblefield et al., 2005: Власова, 1983; Гидрологический режим..., 1977; Кузнецов, 1955).

Four field campaigns were organized from 2011 to 2014. Two of them were the most large-scale, they were held during the summer floods in 2011 and at the low water stage in 2012 to clarify the features of seasonal water runoff and sediment load in the Selenga Basin. Approximately 30 sites within Mongolia and 50 sites in Russia were studied every year in the Selenga River and its tributaries, the discharge and sediment load were measured. Water samples for SSC and turbidity were taken cumulatively in the few gage points across the river width. The samples were filtered through the pre-dried and pre-weighed membrane filter with 0.45 µm pore size during 2 hours under 105 °C, after filtration it was dried during 3 hours under the same temperature. We used portable turbidity meter Hach 2100P for turbidity determining. The water current was measured by a hydrometric propeller ISP-1. The discharges were obtained by acoustic Doppler profiler Rio Grande 600 kHz Teledyne RD Instruments.

Water and suspended sediment budgets were calculated based on the balance method: $\Delta W = W_2 - W_1$, where W_1 – is inflow matter (water or sediment) to the upper reach of the river, W_2 – is outflow material from the reach. If $\Delta W > 0$ in case of water flow, lateral inflow was found; in case of sediment load, the erosion along a river reach was observed. Water losses for evapotranspiration or sediment accumulation were determined for conditions $\Delta W < 0$.

3. Results and Discussion

Analysis of the annual flow water balance for the period from the mid-20th century to the early 21st century has shown that the runoff of the river naturally increases along the Selenga river system from source to sink (Fig. 1).

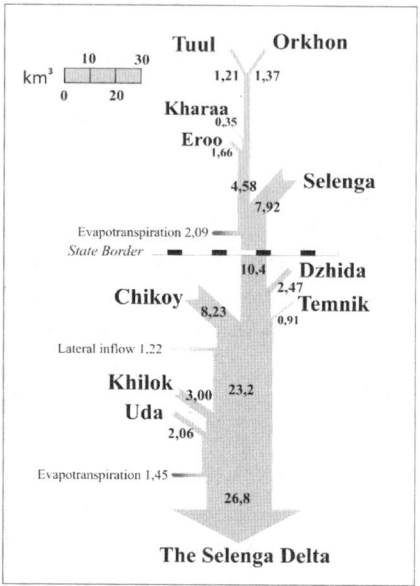

Figure 1. The water budget of the Selenga River Basin from the mid-20th century to the early 21st century

The average annual water flow into Lake Baikal during the study period (1934-2009) was 28.2 km^3. The Orkhon River flow reached 16% from this value, and the total flow from the Mongolian part of the Selenga Basin provided 37%. The most full-flowing tributary of the Russian part is the Chikoy River, its water composed 29% of the total runoff of the Selenga River. The evapotranspiration losses in the river basin estimated at 3.54 km^3 or 13% of the total water flow in the mouth.

Lag time changes depending on the river water regime: during low water season the water from the upper of the Selenga River Basin reaches Lake Baikal for 11-12 days, for floods it flows faster – for 7-8 days (Гармаев, Христофоров, 2010).

Water flow largely determines the sediment load, their synchronous oscillation is usually observed. However, since the 1980s this trend changed, there is a decrease of sediment load due to the degradation of agriculture in the region (Потемкина, 2011).

The data was obtained during the expeditions allows to analyze the sediment load spatial distribution from the source to the mouth in certain phases of the water regime and to study the contribution of flow transport capacity.

Rainfall in July–August 2011 caused an intensification of the catchment and slope erosion, an increase of water flow and transport capacity. The totality of these factors caused the rising of SSC along the river systems of the Selenga Basin. SSC within the river network of the Selenga Basin ranged from 1.43 mg/l (the upstream of the Dzhida River) to 2847 mg/l (the Orkhon River at Kharkhorin) during the flood season. The average SSC for the Mongolian part of the basin was 270 mg/l, which is 15 times higher than the value in Russian part (19.0 mg/l). Along the Orkhon-Selenga river system, where the human impact is minimal, a decrease in SSC from the upstream to the mouth is 160 times (Fig. 2).

Regular decrease in SSC along the rivers during high water phase violated under the influence of industry. SSC of the Tuul River downstream from Ulaanbaatar increased 5-fold compared with the baseline SSC (from 1.68 to 8.36 mg/l), downstream from the Zaamar Goldfield it increased almost three-fold (from 107 to 289 mg/l). Similar processes were studied along the Boroo River and Khangal River. In the first case, SSC increased in 1.6 time from 39.2 to 64.2 mg/l downstream from the Boroo Gold placer mining. Along the Khangal River SSC multiplies 18 times (from 15.2 to 266 mg/l) below a tailings pond of the Erdenet copper-molybdenum mining and processing works.

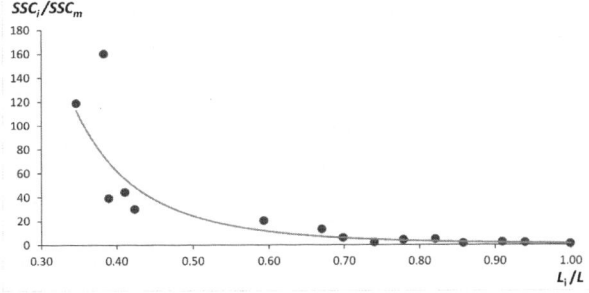

Figure 2. Longitudinal SSC distribution during flood season (July-August 2011): SSC_i, SSC_m – suspended sediment concentration in a site and in the mouth (10.6 mg/l), respectively; L_i – distance from the source: L – the total length of the Orkhon-Selenga river system (1546 km)

SSC gradually decreased due to the energy flow losses from the upstream to downstream of the river network. As a result, there was a process of accumulation of suspended sediment 1145 tons/day in the midstream. The Selenga River received tributaries (Dzhida, Temnik, Chikoy) in the Russian part of the basin, which delivered 775 tons/day of suspended particles because of floods. The accumulation was 2,445 tons/day in the downstream of the river due to reducing transport capacity and decreasing channel slope (Fig. 3).

During the low-flow period (June 2012) the average SSC in the Selenga Basin was 74.5 mg/l, the minimum value was observed in the upstream of the Tuul River (1.68 mg/l), the maximum SSC was 618 mg/l in the upstream of the Khaara River when a local flood occurred. At the upper part of the Selenga Basin SSC was significantly lower compared to the flood season due to lower water flow. Along the Orkhon River SSC increased 5 times from 4.20 mg/l in the upstream to 19.9 mg/l in the mouth due to the channel erosion. For the Orkhon-Selenga River System SSC increased from the source to the mouth in 15 times according to rising water flow. This pattern was changed in local reaches of the river network in connection with an additional input of sediments from the tributaries (Fig. 4).

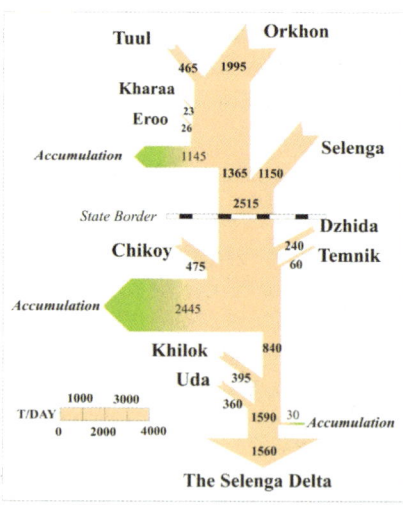

Figure 3. The sediment budget of the Selenga River Basin during flood season (July-August 2011)

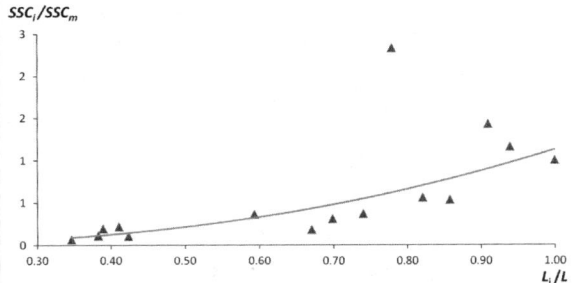

Figure 4. Longitudinal SSC distribution during low water season (June 2012) along the Orkhon-Selenga river system (the designation the same with Fig. 2)

The accumulation of suspended particles 34 tons/day was found at the Orkhon downstream below inflow of the Tuul River with mining areas in the basin and local floods in the Khaar and Eroo Rivers. Despite supplementation of sediment load 636 tons/day by the Dzhida, Temnik and Chikoy Rivers, the sediment starving took place in the lower part of the Selenga River, it caused the erosion of sediments that were deposited earlier in the flood period, and their re-entry into the stream was 428 tons/day. However, about 10% (186 tons/day) of the total sediment load again accumulated before the Selenga delta (Fig. 5). As a result, for the low water phase SSC mainly depends on the vertical sediment fluxes between the flow and alluvium.

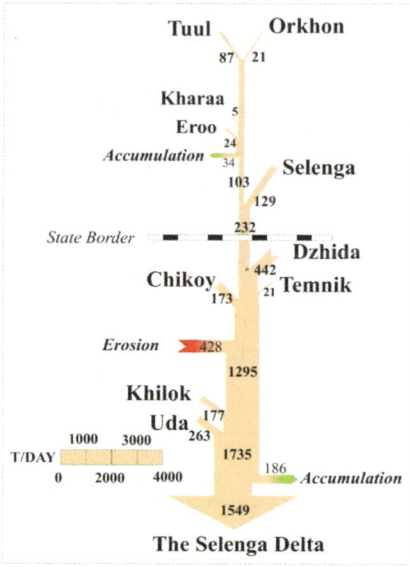

Figure 5. The sediment budget of the Selenga River Basin during low water season (June 2011)

Comparison of the sediment load in the various phases of the water regime to the average annual data for the 1952-2009 (Fig. 6) shows that in general removal of material down the river system dominates within the Russian part of the Selenga Basin. It is caused by intense slope erosion in the spring during snow melt, when the no grass-covered on the surface of the catchment.

The average annual sediment flux through Russian-Mongolian border was 3455 tons/day. For the summer floods transboundary sediment load was 73% of this value and for the low water phase was 7%. According sediment load of tributaries and combine of erosion and deposition of sediment their average annual delivery to the Selenga delta was 5,700 tons/day. During both the low water period 2012 and high water 2011 this value was about 27% of the annual sediment load to the delta: 1,549 and 1,560 tons/day, respectively.

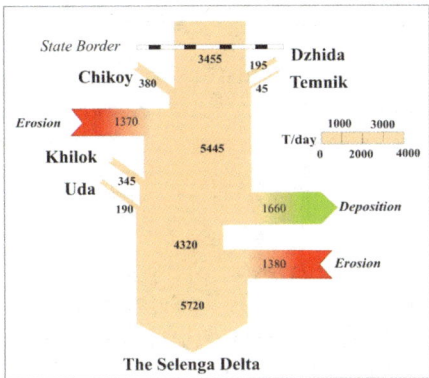

Figure 5. The average annual sediment budget of the Selenga River Basin (1952-2009)

The maximal sediment accumulation along the river network was observed during the peak flood, its value run up to 2445 tons/day for the Selenga River reach from the confluence with the Orkhon River to the Chikoy River.

4. Conclusions

Water flow naturally increased along the Selenga River network, on the Russian-Mongolian border it consisted 37% of the flow at the river mouth. Depending on the water stage the flow differently affected the longitudinal distribution of SSC. In flooding 73% of the average annual sediment load passed the border, in the low water – only 7%.

Longitudinal sediment flux was depend on erosion and transport capacity. SSC increased from source to mouth for low water season in 15 times due to channel erosion. It decreased in 160 times during flood season because of sediment deposition. In consequence of human impact, SSC increased significantly 3-8-fold downstream from the non-ferrous metals mining.

Channel storage determined sediment budget in the downstream part of the Selenga River during high water period. For low water season channel erosion was dominated at the catchment.

Acknowledgement

This study is financially supported by Russian Fund for Basic Research (projects 14-05-31351, 15-05-05515, 15-05-03752) and Russian geographical society grant "Expedition Selenga–Baikal".

References

Chalov S.R., Jarsjö J., Kasimov N.S., Romanchenko A.O., Pietron J., Thorslund J., Belozerova E.V. (2015): Spatio-temporal variation of suspended transport in the Selenga Basin (Mongolia and Russia). Environ. Earth Sci. 73 (2): 663-680.

Karthe D., Kasimov N., Chalov S., Shinkareva G., Malsy M., Menzel L., Theuring P., Hartwig M., Schweitzer C., Hofmann J., Priess J., Lychagin M. (2014): Integrating multi-scale data for the assessment of water availability and quality in the Kharaa - Orkhon - Selenga river system. Geography, environment, sustainability 3 (7): 65-86.

MN-Mining Infrastructure Investment Supp (2013): Available on www.worldbank.org/projects/P118109/mn-mining-infrastructure-investment-supp?lang=en (Accessed on 14th Januarry 2015).

Nadmitov B., Hong S., Kang S.I., Chu J.M., Gomboev B., Janchivdorj L., Lee C.H., Khim J.S. (2015): Large-scale monitoring and assessment of metal contamination in surface water of the Selenga River Basin (2007–2009). Environ. Sci. Pollut. Res. 22: 2856-2867.

Pfeiffer M., Batbayar G., Hofmann J., Siegfried K., Karthe D., Hahn-Tomer S. (2015): Investigating arsenic (As) occurrence and sources in ground, surface, waste and drinking water in northern Mongolia. Environ. Earth Sci. 73: 649-662.

River without Boundaries, Russia urges the World Bank to stop funding projects threatening Lake Baikal (2013): Available on www.transrivers.org/2013/951/ (Accessed on 14th Januarry 2015).

Stubblefield A., Chandra S., Eagan S., Tuvshinjargal D. Davaadorzh G., Gilroy D., Sampson J., Thorne J., Allen B., Hogan Z. (2005): Impacts of Gold

Mining and Land Use Alterations on the Water Quality of Central Mongolian Rivers. Integrated Environmental Assessment and Management 1 (4): 365-373.

Thorslund J., Jarsjö J., Chalov S.R., Belozerova E.V. (2012): Gold mining impact on riverine heavy metal transport in a sparsely monitored region: the upper Lake Baikal Basin case. J. Environ. Monit. 14: 2780-2792.

Withanachchi S.S., Houdret A., Nergui S., Ejarque Gonzalez E., Tsogtbayar A., Ploeger A. (Re)configuration of Water Resources Management in Mongolia (2014): A Critical Geopolitical Analysis. Kassel university press: 42.

Власова Л.К. (1983): Речные наносы бассейна озера Байкал. Новосибирск: Наука: 136. Publication in Russian language. [Vlasova L.K. (1983): The river sediments of the Lake Baikal Basin. Novosibirsk: Nauka: 136.]

Гармаев Е.Ж., Христофоров А.В. (2010): Водные ресурсы рек бассейна озера Байкал: основы их использования и охраны. Новосибирск: Академическое издательство «ГЕО»: 227. Publication in Russian language. [Garmaev E.J., Khristovorov A.V. (2010): Water Resources of the Rivers of the Lake Baikal Basin: Basics of Their Use and Protection. Novosibirsk: Academic Press "Geo": 227.]

Гидрологический режим рек бассейна р. Селенги и методы его расчета (1977). Ленинград: Гидрометеоиздат, 1977: 236. Publication in Russian language. [The hydrological regime of the Selenga River Basin and its calculation methods (1977). Leningrad: Gidrometeoizdat, 1977: 236.]

Гомбоев Б.О., Зомонова Э.М., Зандакова А.Б. (2014): Проблемы управления водными ресурсами трансграничного бассейна реки Селенги. Проблемный анализ и государственно-управленческое проектирование 5 (37), том 7: 78–87. Publication in Russian language. [Gomboev B.O., Zomonova E.M., Zandakova A.B. (2014): Problems in the management of water resources of the transboundary Selenga River Basin. Problem analysis and public administration projection 5 (37), Volume 7: 78-87.]

Кузнецов Н.Т. (1955): Основные закономерности режима рек Монгольской Народной Республики. Москва: Издательство АН СССР: 104. Publication in Russian language. [Kuznetsov N.T. (1955): The basic laws of the river regime of the Mongolian People's Republic. Moscow: Publishing House of the USSR Academy: 104.]

Потемкина Т.Г. (2011): Тенденции формирования стока наносов основных притоков озера Байкал в XX и начале XXI столетия. Метеорология и гидрология 12: 63–71. Publication in Russian language. [Potemkina T.G. (2011): Trends in the formation of sediment load major tributaries of Lake Baikal in the XX century and the beginning of XXI century. Meteorology and Hydrology 12: 63-71.]

Morphological analysis of the upper reaches of the Kukuy Canyon derived from shallow bathymetry

Nicolas Le Dantec[1,2], Nathalie Babonneau[1], Marcaurélio Franzetti[1], Christophe Delacourt[1], Yosef Akhtman[3], Alexander Ayurzhanaev[4], Pascal Le Roy[1]

[1] Université Européenne de Bretagne Occidentale, UMR-6538 Domaines Océaniques, UBO-CNRS, Plouzané, France
[2] Centre d'Etudes et d'Expertise sur les Risques, l'Environnement, la Mobilité et l'Aménagement, DTechEMF, Margny Lès Compiègne, France
[3] Geodetic Engineering Laboratory, Ecole Polytechnique Fédérale de Lausanne, Lausanne, Switzerland
[4] Baikal Institute of Nature Management, Siberian Branch of the Russian Academy of Sciences, Ulan-Ude, Republic of Buryatia, Russian Federation

Abstract

We present preliminary results on the morphology of the upper reaches of the Kukuy Canyon and Selenga shelf in front of Proval Bay (Lake Baikal), derived from newly acquired, high-resolution bathymetry. Numerous and varied erosional and transport features provide an interpretation framework for source to sink transfer and gravity flow processes in this shallow and active tectonic environment, suggesting on-going gravity instabilities and sediment-laden flows. Scarps in the canyon head are likely signatures of retrogressive incision of the western tributary and eastward lateral migration of the western tributary, the latter coming within about 1 km of the shoreline. Immature gullies incising the upper-slope feedings of the Kukuy Canyon indicate gravity flows with low erosional power. Large arcuate scarps on the break of the narrow shelf east of Proval Bay reveal gravity instabilities. The morphological connection between the Selenga Delta and the Kukuy Canyon suggests a direct pathway for fluvial sediment focused through breaches in the Sakhalin sand shoal, with likely occurrence of hyperpycnal flows into canyons heads during high sediment discharges. The neotectonic activity affects both the accommodation space around the prograding delta via earthquake-induced subsidence of coastal areas, and the location of incisions through slope instability triggering. Subsequent surveys allowing diachronic analysis would help determining the influence of tectonic and climatic factors controlling sediment transfer across the land-lake continuum and interpreting the morphological signature of the associated gravity processes shaping the delta and surrounding shelf and canyons.

1. Introduction

Sediment transfer from land to deep basin is a major process in the formation of landforms and sub-aquatic morphologies like submarine canyons. It mainly occurs during extreme climatic or geological events (floods, earthquakes...), supplying large quantities of sedimentary material to basins during brief and repetitive episodes (Korup, 2012). In deep environments, gravity flows as turbidity currents are mostly responsible for canyons incision in the continuity of the main rivers. The triggering of turbidity currents remains a topical research question in marine environments (Piper and Normark, 2009). Scientific issues include the continuity at sea of hyper-concentrated stream flows generating hyperpycnal flows into canyons, mass wasting and re-suspension of sediment deposits under the influence of oceanographic factors (waves, currents). The interface between fluvial and submarine sedimentary systems is a key area to understand the morphological evolution and transformation associated with sediment transfer processes. In recent years, high-resolution bathymetric data acquired with latest-generation multibeam technologies in shallow-water environments provides advances in outlining detailed morphological features, and in understanding the sedimentary processes at canyon heads (Smith *et al.*, 2007; Yoshikawa and Nemoto, 2010; Casalbore *et al.*, 2011; Lastras *et al.*, 2011; Babonneau *et al.*, 2013).

Lake Baikal in Siberia is among the best examples of a large tectonic lake. It occupies the central part of the presently still active Baikal Rift Zone. It is the world's largest and deepest freshwater lake. Morphologically, it is subdivided into three deep basins, the South, Central and North Baikal Basins, which are separated by inter-basin highs: the Selenga Delta accommodation zone (or Buguldeika Saddle) and the Academician Ridge accommodation zone. The development of the Baikal basin results from earthquakes, which are accompanied by considerable vertical displacements of the lake bottom, collapses and landslides in the coastal zone (Shchetnikov *et al.*, 2012). Contrary to the pronounced morphological structure (tectonic escarpments) observed on the northwestern side, the more gently sloping eastern side of the rift shows major lithospheric extension associated with rearrangement of blocks in the upper lithosphere slab.

The Selenga River is a major river in Asia and forms a large delta on the southeast shoreline of Lake Baïkal. The mean discharge of the Selenga River into Lake Baikal in winter is approximately 100 $m^3.s^{-1}$, which increases to 1,700 $m^3.s^{-1}$ in spring as the snow melts. The Selenga Delta consists of a multibranch fan-like channel structure. The northern shelf of the delta is incised by a deep NE-SW oriented canyon (the Kukuy sub-aquatic canyon) feeding a large turbiditic fan. The morphology of this fan is characterized by a network of sinuous turbidite channels, as observed in deep-sea fans, which suggest active and frequent sediment transfer from the deltaic channels to the deep lake via turbidity currents or hyperpycnal flows into the canyon.

Proval Bay, to the East of the Selenga Delta, was formed by a large seismic dislocation event: the 1862 earthquake caused subsidence (up to 7-8 m) of the former Tsagan steppe below the water surface. This seismotectonic phenomenon constitutes an example of scaled subsidence of crustal blocks in coastal zones (Shchetnikov *et al.*, 2012). On the south and the east, the littoral zone of the bay is controlled by the Delta fault, which has a defined scarp up to 10-12 m high. Comparison of old charts with modern maps shows that the boundary of the Selenga River delta has shifted considerably eastward (Vologina *et al.*, 2010) due to intense progradation from the Selenga River delta plain (66 mm/yr). Sedimentation rates vary greatly across Proval Bay depending on proximity to the Selenga River. Seismotectonics thus plays a role of buffer mechanism in the sediment transfer through the land-sea interface, as subsidence events generate accommodation space. Sediments prograde into the recently formed coastal embayments instead of being delivered directly to the Baikal basin through the canyon and shelf break. Before the rise of water level when the Irkutsk hydropower station of Angara dam was put into operation, Proval Bay was separated from Lake Baikal by the oblong Sakhalin island.

The remnants of this sand ribbon still exist as an underwater sand shoal consisting of numerous bars bordering most of the bay. The entire system of underwater bars is active, with a general eastward and offshore drift on the order of 50 m/yr (Rogozin, 1993).

We have acquired high-resolution bathymetry in the shallower section of the lake in front of Proval Bay, on the upper reaches of the Kukuy Canyon. Here, we are presenting preliminary results on the morphology of the canyon and shallow shelf. We describe erosional features and infer sediment transport pathways. The main scientific issues are : (1) to identify the potential activity of turbidity currents initiated in the canyon heads by the observation of active bedforms and fresh erosional structures ; (2) to understand the continuity of the sedimentary processes between the delta and the canyons (morphological continuity, transformation of the flows, sediment instabilities…) and (3) to determine the influence of tectonic factors in the triggering of gravity instabilities, and in the control of the location of morphological incisions. This initial study highlights the interest in extending the survey to cover the entire shelf in front of the delta, possibly with repeated surveys to develop a multi temporal approach (Franzetti *et al.*, 2013, Babonneau *et al.*, 2013).

2. Data and methods

A high-resolution bathymetric dataset of the Northern Selenga Delta was collected in August 2014 during the SELENGA 2014 survey, using a shallow-water multibeam echosounder Kongsberg EM3002. This high-resolution 300 kHz multibeam system was temporarily set up on a lake survey boat, which allowed acquisition of bathymetric data in water depths ranging from 2 m up to 250 m with a vertical resolution of up to 10 cm. Sound velocity profiles were carried out twice a day in the vicinity to correct sound refraction errors. Multibeam soundings were edited with QINSy software to provide a 4-m grid DTM. Figure 1 shows the surveyed area.

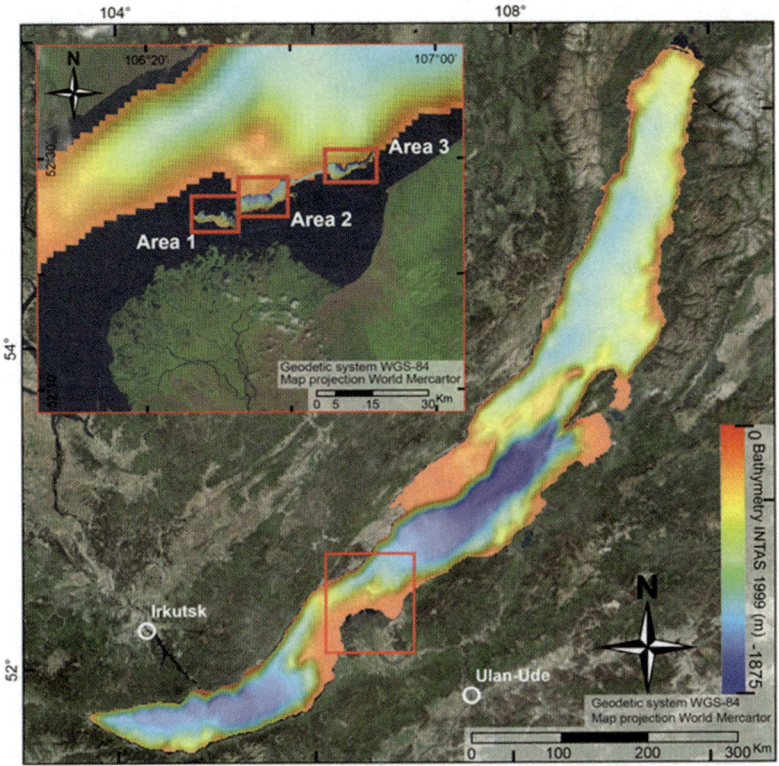

Figure 1: Bathymetric map of Lake Baikal (INTAS, 2002; Sherstyankin et al., 2006). Insert: Location map showing the Selenga Delta and the study area (3 boxes indicating the new high-resolution bathymetric data).

3. Results

The derived bathymetry is decomposed in 3 areas for the presentation of the results (Figure 2). Survey planning was devised to focus on the upper reaches of the canyon, off the Western side of Proval Bay, and on the shelf break of the Eastern side of the bay. In the first of the 2 zones, the delta shows significant focusing of the sediment discharge through the sand bar, which is visible on satellite images (Figure 2). The head of the canyon exhibits a bifurcation, with the main branch turning toward the shore and coming within about 1 km of the shoreline. The second branch is mainly in the same orientation as the canyon itself but it is much less deeply incised. In the results description, the canyon head will constitute area 1. Near the shelf break, the Kukuy Canyon exhibits a sharp lateral jog, where the orientation of the thalweg abruptly turns at 90° and back. The section of the canyon up-canyon of this lateral shift and down-canyon of the canyon head runs nearly parallel to the shelf break. Along this section, secondary channels incised along the upper shelf feed the main canyon beyond the head. This sector will be area 2. While the lake has a mostly narrow shelf (5 km wide on the South shore and 10-15 km wide on the North shore), in the Selenga accommodation zone in front of the delta, the shelf extends over 30 km into the lake. Data collected off the Eastern side of Proval Bay, where the shelf becomes much narrower after a sharp inshore turn of the shelf break (300m isobath), is presented as area 3.

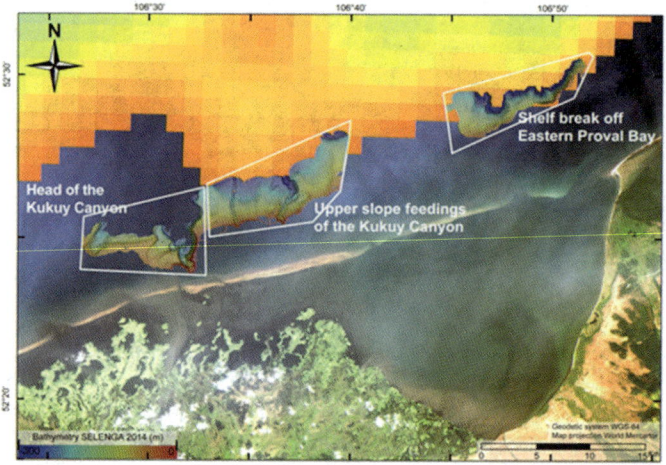

Figure 2: High-resolution bathymetric map at the mouth of the Selenga delta and Proval Bay (Satellite Image: Landsat 8 from August 18th, 2014).

Area 1: Head of the Kukuy Canyon

Of the two tributaries composing the head of the Kukuy Canyon, the western tributary canyon is slightly asymmetric and bounded by a steep and rectilinear left side in the continuity of the canyon side. The NE-SW rectilinear orientation suggests a strong morphological control by a major fault, matching with the main rift structures. The canyon head is characterized by several rounded scarps at less than 50 m of water depth, which can be interpreted as local slope instabilities generating a retrogressive incision of this canyon head. The canyon floor is about 1 km wide and shows irregular bedforms and erosional features. It is not draped by muddy sediment but shows relatively fresh sedimentary features indicating a recent activity of gravity processes.

The eastern tributary canyon is the most incised. Its morphology is highly asymmetric with a steep eastern side oriented N-S and an arcuate morphology. This arcuate morphology is consistent with the sinuous path of the thalweg in the deeper part of the canyon. The presence of current and abandoned meanders along the thalweg indicates a mature canyon incised and active for a long time. The morphology of the western side of the tributary canyon shows several rounded erosional structures suggesting the lateral migration of the canyon head from West to East. The present canyon head shows very fresh scarps initiated at less than 20 m of water depth. The longitudinal depth profile in the canyon axis shows a succession of small steps suggesting active processes of slope instabilities.

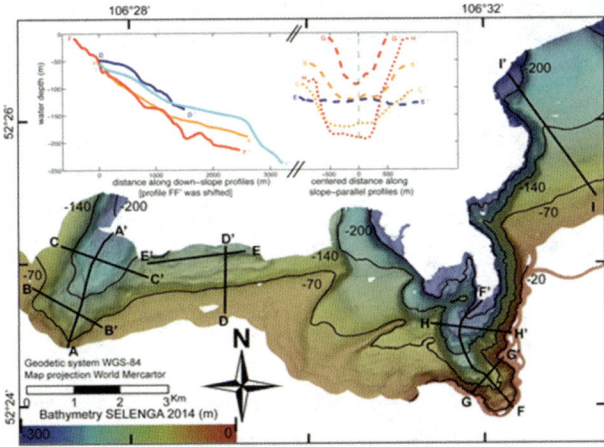

Figure 3: Zoom of the bathymetric map focused on Area 1, western feeding of the Kukuy Canyon (isobaths at 20, 50, 100 and 200 m water depth and location of the bathymetric profiles). Insert: Bathymetric profiles.

Area 2: Upper slope feedings of the Kukuy Canyon

The central area also corresponds to a sector with smaller tributary canyons and gullies incising the upper slope and feeding the Kukuy Canyon. In this area, two small canyons are distinguished with incision about 25 m deep and width about 200-300 m. The sinuosity of these small canyons is low and suggests more immature morphology. Other erosional features are visible. They are straight and their incisions are superficial, corresponding to immature gullies suggesting gravity flows with low erosional power.

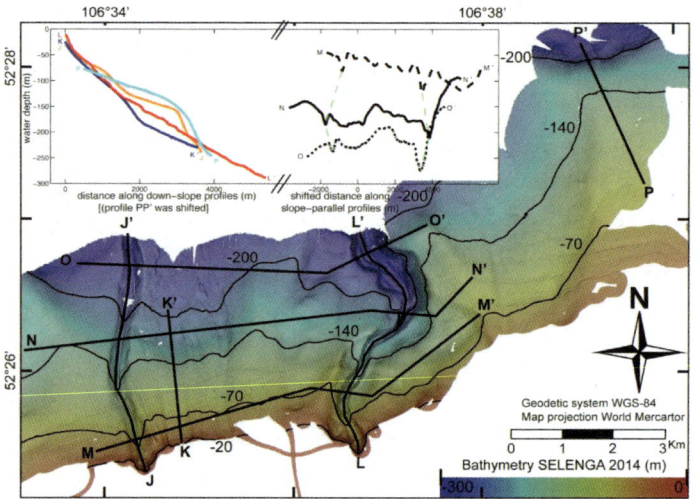

Figure 4: Zoom of the bathymetric map focused on Area 2, eastern feeding of the Kukuy Canyon (isobaths profiles location as in Figure 3). Insert: Bathymetric profiles.

Area 3: Shelf break off Eastern Proval Bay

Shallow bathymetry illustrates the morphology of the shelf break and the upper slope at the eastern extremity of the delta. The eastern area is not directly linked to the Kukuy Canyon. In this area, the submarine slope is steeper east-ward and shows large arcuate scarps. Such morphology probably corresponds to gravity instabilities of the shelf break, which contribute to mass transport through the canyon, generating slump, debris flow or turbidity flow. Earth-quakes as well as sediment overload are likely trigger mechanism for slope failure. The absence of morphological evidence of sediment-laden, bottom flows in the shallow bathymetry between the delta and the shelf break, contrary to the 2nd area, can be explained by the earthquake-induced creation of Proval Bay offering accommodation space for the prograding delta.

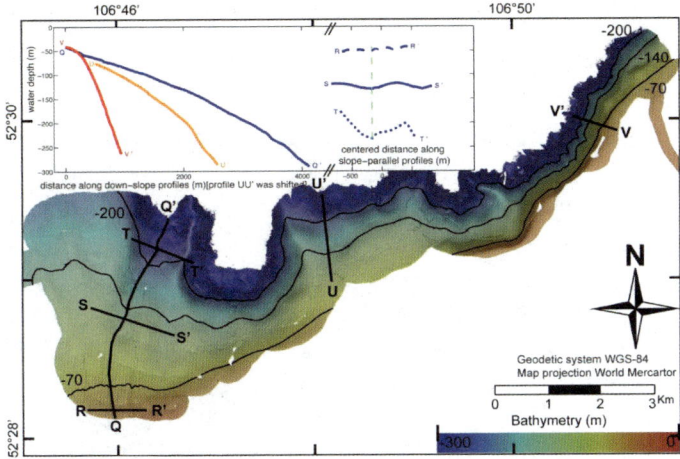

Figure 5: Zoom of the bathymetric map focused on Area 3, Shelf break and upper slope of the eastern part of the delta (isobaths profiles location as in Figure 3). Insert: Bathymetric profiles.

4. Conclusions and perspectives

The interpretation of shallow bathymetric data aims to show the detailed struc-ture of the upper reaches of the Kukuy Canyon. Preliminary results provide new elements on the issues of source to sink transfer and gravity flow pro-cesses in this shallow and active tectonic environment. The number and vari-eties of sedimentary features and the recently-generated morphologies (scarp, bedforms ...) observed in shallow water suggest active gravity instabilities and gravity flows as turbidity currents. The morphological connection and close proximity between the delta channels breaching the sand bar and the upper reaches of the Kukuy Canyon suggest a direct pathway for fluvial sediment. It is likely that high sediment discharge volume of the delta during snow melt periods generates hyperpycnal flows feeding the Kukuy canyons heads. More-over, the active seismic setting both affects the accommodation space around the prograding delta, with morphological changes of the lake basin through earthquake-induced subsidence, and is also a very likely predisposing factor for slope instability triggering, with a strong tectonic control on the location of incisions and scarp failures by rift faults. Multitemporal bathymetric surveys will be carried out in the next years, providing new measurements in order to better understand the relationships between variation of sediment discharge and gravity processes, and the links between shallow and deep gravity processes.

Acknowledgments

Authors thank the Pôle Image (IUEM, Brest) for technical support (multibeam echosounder), the staff of the research center in Istomino, the crew on the boat, all the collaborators on the Leman-Baïkal project and Leonid Byzov (Irkutsk University), with a special note for Mideg Dugarova who assisted us as a translator. The project benefits from the financial support of Critex Equipex (ANR) and LabEx MER (ANR-10-LABEX-19) projects, the Ministry of the French Foreign Affairs and the EU through the FP7 project IQmulus (FP7_ICT_2011_318787).

References

Babonneau N., Delacourt C., Cancouet R., Sisavath E., Bachelery P., Mazuel A., Jorry S.J., Deschamps A., Ammann J. and Villeneuve N. (2013), Direct sediment transfer from land to deep-sea: insights from new shallow-marine multibeam data at La Réunion Island, Marine Geology 346, 47-57

Casalbore, D., Chiocci, F., Scarascia Mugnozza, G., Tommasi, P. and Sposato, A., 2011. Flash-flood hyperpycnal flows generating shallow-water landslides at Fiumara mouths in Western Messina Strait (Italy). Marine Geophysical Research 32(1), 257-271. doi: 10.1007/s11001-011-9128-y.

De Batist M.A., Canals M., Sherstyankin P., Alekseev S. and the INTAS Project 99-1669 Team (2002) A new bathymetric map of Lake Baikal.

Franzetti M., Le Roy P., Delacourt C., Garlan T., Cancouët R., Sukhovich A and Deschamps A. (2013), Giant dune morphologies and dynamics in a deep continental shelf environment: example of the Banc du Four, Western Brittany, France, Marine Geology 346, 17-30

INTAS Project 99-1669 Team, October 2002. A New Bathymetric Map of Lake Baikal.   http:/allserv.ugent.be/   ~mdbatist/intas/intas.htm, http:/www.lin.irk.ru/intas/ index.htm.

Korup O., 2012. Earth's portfolio of extreme sediment transport events. Earth-Science Reviews 112, 115-125. doi:10.1016/j.earscirev.2012.02.006.

Lastras, G., Canals, M., Amblas, D., Lavoie C., Church, I., De Mol, B., Duran, R., Calafat, A.M., Hughes-Clarcke, J.E., Smith, C.J., Heussner, S., 2011. Understanding sediment dynamics of large submarine valleys from sea-floor data: Blanes and la Fonera canyons, northwestern Mediterranean Sea, Marine Geology, 280, 20-39. doi: 10.1016/j.margeo.2010.11.005.

Naudts L., Khlystov O., Khabuev A., Seminskiy I., Casier R., Cuylaerts M., Synaeve J., Vlamynck N., De Batist M.A. and Grachev M.A. (2009) Newly collected Multibeam Swath Bathymetry data herald a new phase in gas-hydrate research on Lake Baikal. AGU Fall Meeting 2009.

Piper D.J.W., Normark, W.R., 2009. Processes that initiate turbidity currents and their influence on turbidites: A marine geology perspective. Journal of Sedimentary Research 79, 347-262. doi: 10.2110/jsr.2009.046.

Rogozin, A.A., 1993. Littoral Zone of the Baikal and Khubsugul. Nauka, Novo-sibirsk. (in Russian).

Shchetnikov A.A., Radziminovich Y.B., Vologina E.G., Ufimtsev G.F., The formation of Proval Bay as an episode in the development of the Baikal rift basin: A case study, 2012. Geomorphology 177-178(1), pp. 1-16 doi:10.1016/j.geomorph.2012.07.023

Sherstyankin P.P., Alekseev S.P., Abramov A.M., Stavrov K.G., De Batist M., Hus R., Canals M. and Casamor J.L. (2006), Computer-Based Bathymetric Map of Lake Baikal, Doklady Earth Sciences 408 (4), 564-569.

Smith, D.P., Kvitek, R., Iamietro, P.J., Wong, K., 2007. Twenty-nine months of geomorphic change in upper Monterey Canyon (2002-2005). Marine Geology 236, 79-94. doi: 10.1016/j.margeo.2006.09.024.

Vologina, E.G., Kalugin, I.A., Osukhovskaya, Yu.N., Sturm, M., Ignatova, N.V., Radziminovich, Ya.B., Dar'in, A.V., Kuz'min, M.I., 2010. Sedimentation in Proval Bay (Lake Baikal) after earthquake-induced subsidence of part of the Selenga River delta. Russian Geology and Geophysics 51, 1275–1284. http://dx.doi.org/ 10.1016/j.rgg.2010.11.008.

Yoshikawa, S., Nemoto, K., 2010. Seasonal variations of sediment transport to a canyon and coastal erosion along the Shimizu coast, Suruga Bay, Japan. Marine Geology 271,165-176.

IV. State of aquatic and terrestrial ecosystems

Geo-ecological Issues in the Selenge River Basin Catchment

Enkh-Amgalan. S.[1]**, Dorjgotov D.**[1]**, Oyungerel J**[1]**, Enkh-Taivan D.** [1]**, Batkhishig O.**[1]

Institute of Geography, Mongolian Academy of Science
Email: amgalan69@yahoo.com

Abstract

The Selenge basin belongs to the central part of Mongolia and it is a core region of considerable deposits of mineral resources of Mongolia, infrastructure, roads, and arable lands, it can be deemed as a key region of the country's development. Selenga Basin is located in a very unique landscape composed of various combinations of natural conditions. It moves outside the basin area transition in the boundary regions in the unique natural ecosystems and plays an important role to maintain the ecological balance in Mongolian.

The water, soil and air are polluted due to technogen, in particular the level of environment pollution to be reach critical amount near to mining industry and settled area of soums, aimags, Erdenet, Darkhan and Ulaanbaatar city. Big industrial centers with dense population, (Ulaanbaatar, Darkhan, Erdenet), certain local mining enterprises, leather and wool processing light industries cause negative impacts to the environment by deteriorating nature's manner and polluting water, soil and air. Especially, due to mountain mining activities, small rivers and streams are getting wiped out thus a problem of water lack is becoming tensional. Consequently, the issue to use water reserve for production has been more vital in Mongolia. There is too much air pollution in Ulaanbaatar city, especially, in wintertime, because of using plenty of brown coal.

Keywords – Socio-economic potential, arable land, land degradation, soil resources, landscape change, morphogenetic types

Socio-Economic Potentials of the Selenge River Basin

Economically, the Selenge basin belongs to the central part of Mongolia and accounts for 20.3%of the total population of the country, 19.7% of families, 20.8% of industrial product sales, 68.9% of arable lands and 26.1% of total livestock respectively (Fig 1).

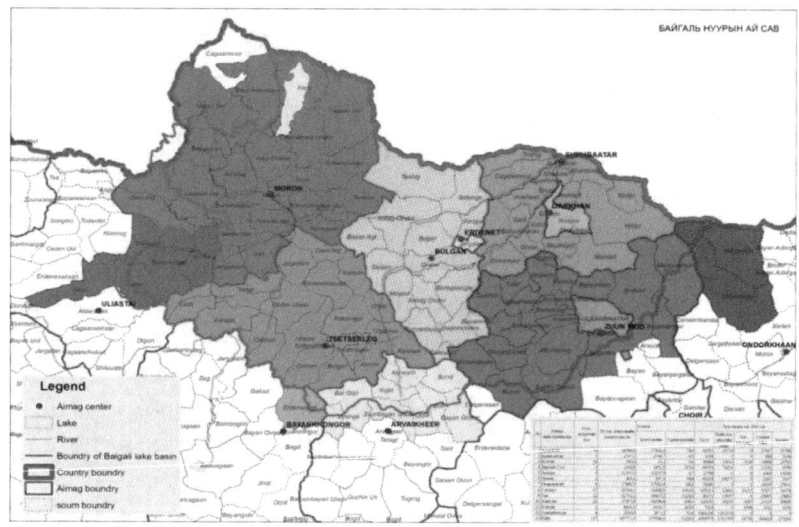

Fig1. Map of Selenge River Basin

Migration of the people has been on the increase recently, to the cities and their neighboring areas [1]. The Selenge river basin totally includes 21 specially protected areas, and covers 592,2 thousand hectares of lands thus accounting for 37,2% of the territory of Mongolia and 11,3% of the basins. Selenge basin-4 protected areas, 4 natural reserves, 10 natural complexes, and 3 natural sightseeing. 1. Special protected areas, 2.Natural complexes, 3.Natural reserves, 4.Natural sightseeing, 5.Locally protected areas [2].

In Mongolia, it started to mine gold in the vicinities of Yeruu and Boroo river basins, in Khuvsgul area from beginning of 20th century. Such deposits as Boroo and Bumbat near Zaamar region are considered having big reserves and some of plate layers comprise 10g/tn of gold content. The former gives 5 tons and the latter does 1.5 tons of gold a year. Shariin gol brown coal deposit has been exploited since 1961 by means of open cast mine and as of 2010, around 1 million tons of coal are mined. There are coal mines like Ulaan-Ovoo Brown coal deposit with high heat in Selenge province, Julchig and Nuurstein Am in Khuvsgul province. Ereen and Saikhan-Ovoo coal deposits in Bulgan province where coals are mined in little amount respectively. Since 1912, Nalaikh coal mine has been exploited and 600-800 thousand tons of coal were mined a year that used to provide thermo power stations of Ulaanbaatar city with coal thus it was a sole source of fuel. But it was closed now because of security problem [3].

As for economic zone, it comprises 189 soums and 881 bhags of 11 provinces and has a total population of 985 thousand. There are three big cities such as Ulaanbaatar, Darkhan and Erdenet in the region.

Table 1. Economic and Population parameters

№	Provinces name	Soum	Baga	Territory /thou.km 2/	Population 2013 (thou. person)			
					Total	Aimag center	Soum center	Count ryside
1	Zavkhan	24	114	82.4	69.7	15.5	14.6	39.6
2	Arkhangai	19	99	55.3	92.5	20.7	17.9	53.7
3	Bayankhongor	20	100	116	82.2	27.1	11.7	43
4	Bulgan	16	72	48.7	58.8	12	18	28.8
5	Orkhon	2	19	0.8	91.1	85.3	2.2	3.6
6	Khovsgul	24	121	100.6	123	38.8	27	60.7
7	Darkhan-Uul	4	24	3.3	93.5	74.4	12.5	6.6
8	Selenge	17	49	41.2	105.2	22.5	66.3	16.4
9	Tov	27	97	74	88.4	15.4	33	40.2
10	Khentii	17	83	80.3	69.2	17.5	22	25
11	Uvurkhangai	19	103	62.9	111.8	28.5	33	50.2
12	Ulaanbaatar	9		4.7	1287.1			
	Total	189	881	745.8	2340.6	447.6	205.8	398

2013, 25.0 thousand head of livestock have been counted in the provinces of Selenge basin thus it accounts for 40% of the total livestock of the country. The most percentage of livestock is in Tuv, Uvurkhangai and Bayankhongor provinces. The least percentage of livestock is in Orkhon and Darkhan-Uul provinces.

№	Provinces	2013					
		Horse	Cow	Sheep	Goat	Camel	Total
1	Zavkhan	130,128	115,283	128,001	1,007,298	6,289	1,386,999
2	Arkhangai	268,152	427,151	1,944,074	1,131,782	1,107	3,772,266
3	Bayankhongor	108,041	145,728	835,421	1,860,648	37,226	2,987,064
4	Bulgan	225,681	221,005	1,449,566	859,808	1,125	2,757,185
5	Orkhon	11,569	21,629	79,864	65,749	145	178,956
6	Khovsgul	165,535	316,835	1,722,754	140,214	1,915	2,347,253
7	Darkhan-Uul	12,812	3,924	144,924	81,892	185	243,737
8	Selenge	71,467	171,477	624,195	446,557	5	1,313,701
9	Tov	283,635	239,417	1,815,912	1,231,286	2,738	3,572,988
10	Khentii	195,587	20,718	1,408,743	1,060,238	3,586	2,688,872
11	Uvurkhangai	205,881	144,926	1,541,304	1,574,192	1,974	3,468,277
12	Ulaanbaatar	30,202	64,611	134,616	99,540	12	328,981
	Total	1,708,690	1,892,704	11,829,374	9,559,204	56,307	25,046,279

Table 2. Livestock Husbandry Situation 2013 [1]

The Morphogenetic Types of Reliefs, Formation and Landscape Change of Selenge Basin

Selenga Basin is located in a very unique landscape composed of various combinations of natural conditions. It moves outside the basin area transition in the
boundary regions in the unique natural ecosystems and plays an important role
to maintain the ecological balance in Mongolian.

In other words, here is Asia type, on the other hand, ingressive occurrence south Siberia Natural Park. The surface of the external and internal process of the Earth's crust as a combination of service and the geological era
long- term development agenda has grown as a result of tectonic erosion processes [4]. Selenge river basin included geomorphological great region covers
a large territory such relief as wide open steppe, intermountain depression,
between hills valleys, foothills, hillocks, small, medium and high mountains. In
the Selenge river basin include northern part of the Siberian great region, and
region of Huvsgul, of Murun, western and eastern region of Huvsgul, southern
part of the Central Asian state geomorphological creat region and region of
Orkhon-Selenge and north, north western of Khangay region, respectively. Distribution of relief for branch Khentii Mountains and north-east of region of
Khentii Mountain sector, Baga Khentii Mountains include ancient glacial fea-
tured strong dissected, steep slopes high mountains to the medium dissected
middle Mountain, Baga Khentii Mountains and it's along the Branch Mountains,

east part of Buren and Buteel Mountains, basin of Orkhon- Selenge, Low Mountains, hummocky, erosion and upland plain combination with surface. Therefore, morphogenetic types of relief have different structures. The relief, distributed in Selenge river basin divided in to following 4 main types: 1.Tectonic erosion, 2.Volcanic, 3.Erosion, 4. Accumulation [5].

Global warming and our country's natural resources especially production mineral mining affect to change structure of relief and landscape type. Selenge River Basin represented Selenge province, during the past 21 years, the land changes of the area can be divided into two phases between 1990, 2011. We did a research work on the progress of landscape-land cover changes in the study area from 1990 through 2011 using landscape studies data and satellite data. Landscape and land cover classification of Selenge province classified basic types level 7 these are forest, forest steppe, meadow steppe, wetland steppe, lakes, and affected by human activities landscape (urban land, cropland and affected by mining activities). As a result of the research, we found the following landscape-land cover changes in the area (Table 3, Fig 2, 3).

№	Type	Total area, 1990 years		Total area, 2011 years		Landscape-land cover change (1990-2011 years)	
		hectare	%	hectare	%	hectare	%
1	Forest	1623817	39.4	1411120	34.2	-212697	-13.1
2	Forest steppe	215150	5.2	315200	7.6	100050	46.5
3	Meadow steppe	51513	1.2	45150	1.1	-6363	-12.4
4	Wetland steppe	130970	3.2	115270	2.8	-15700	-12.0
5	Steppe	1754280	42.5	1863970	45.2	109689.6	6.3
6	Lakes	34021.6	0.8	34647	0.8	625.4	1.8
	Affected by human activities landscape						
7	Urban land	21354	0.5	21945	0.5	591	2.8
	Cropland	293570	7.1	311721	7.6	18151	6.2
	Affected by mining activities	324	0.008	5977	0.145	5653	17.4 increased again
	The total territory: 4125000 ra						

Table 3. Landscape- land cover classification its change, Selenge province (1990-2011 years)

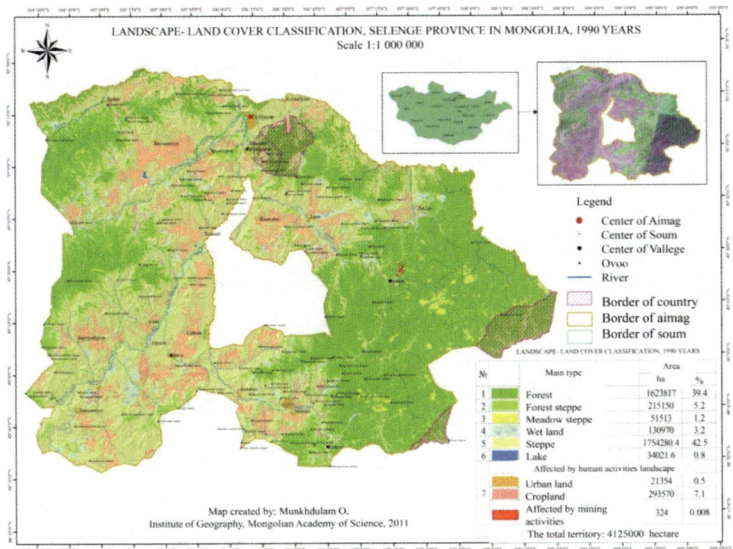

Figure 2. Land Cover Classification 1990

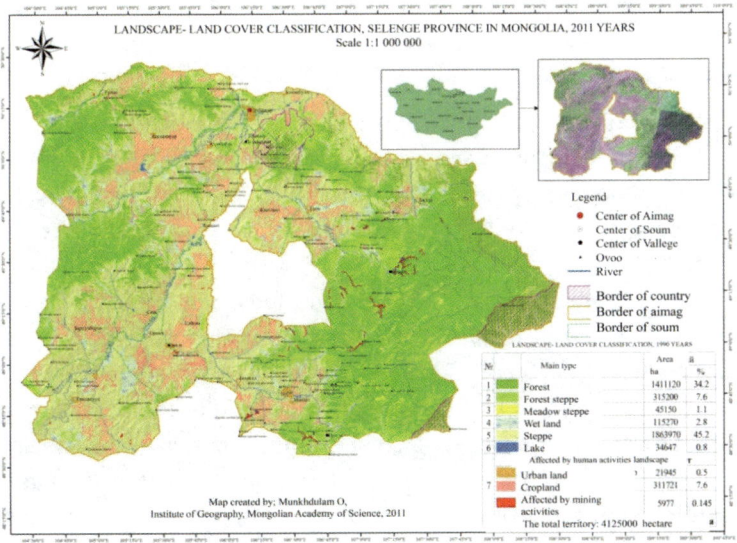

Figure 3. Land Cover Classification 2011

During, 1990-2011 landscape and land cover had changed in the province [6]:

- Forest area decreased by 212.7 thousands hectare.
- Forest steppe area increased by 100.0 thousands hectare.
- Meadow steppe area decreased by 6.4 thousands hectare.
- Wetland steppe area decreased by 15.7 thousands hectare.
- Steppe area increased by 109.7 thousands hectare.
- Lakes area increased by 0.7 thousands hectare.
- Urban land area increased by 21.4 thousands hectare.
- Cropland area increased by 18.1 thousands hectare.
- Affected by mining activities area increased by 5.9 thousands hectare

Global climate warming, mineral extraction chaos and excessive grazing pressure, cutting of forests and other natural and human activity due to its nature as a result of service associated with complex and changing ecosystem balance is achieved. In other words, an increase in the scope of use of land-based resources, environmental, economic, social and environmental issues can be complex. Into areas due to excessive pressure and degrade, a negative impact increase has trend change in the local environment.

Geo-Ecological Issues of the Selenge River Basin

In the basin of the Selenge River the climate change and aridity have become intensively in the comparison of the other regions, Mongolia. For instance, above said region the annual average air temperature has been increased by 2.4^0C (in the other regions by 2.1^0C), while the annual precipitation has decreased by 18% [7].

Because the entire Selenge river basin is located mountain-forest-steppe zone, for the soil resources mountain taiga and mountain steppe soils cover the main part of the basin. According to the small-scale soil map, 35% (10.2M ha) of the total area of basin cover mountain forest and taiga soil, including mountain taiga cryomorphic soil and mountain derno-taiga soil, mountain taiga podzolic soil, and mountain forest-dark colored soil are dominated and forest slightly podzolic sandy soil involve very much area. The mountain chernozem and mountain kastanozem soils occupy 37.4% (10.9M ha) of the total area while high mountain meadow and meadow steppe soil, mountain tundra soil all together take 7.2% (2.1M ha) of the Selenge river basin. In the depressions between mountains, river valleys, terraces and levels mostly become spread chernozem, kastanozem, meadowish kastanozem soils (15.8% or 4.6M ha) and meadow-boggy soil and floodplain soil, water surface take the other parts (4.3% or 1.2 million hectare) [8].

In the exploitation of the soil, pasture and grazing land are mostly dominated and forestry and arable land occupy a considerable part. Due to developing settlements, mining, industry and road network, there tendency to be decreased agricultural land. On the other hand, because of human activity, technogen process, facts such as erosion, degradation, pollution and losing of the fertility of the soil, the condition of the environment is deteriorating [9].

About 80% of total area of arable land and 70% of the woodlot of the Mongolia is located in the basin of the Selenge River. So, it is evident that lost and degradation of the arable land and forest soil, today have become a serious problem. In 2002, arable land area was 756T ha, but in 2012 it was increased to 1031T ha, and considerably land was additionally tilled. In 2012, there was abandoned 292T ha of land and this land seriously broke down and impossible to plant [9]

Loosing and decline fertility of the agricultural soil is connected first of all with the passes of the surface soil got off by wind erosion. Today almost half of total arable land in some degree is subject to break up and fertility of soil is decreased. The losses of soil lighter particle and soil organic carbon due to wind erosion in the highly eroded cultivated area are 131-195 t/ha and 40-60%, respectively. In the moderately eroded cultivated area, there are 70-110 t/ha and 15-30%, respectively rather then 24-54 t/ha and 5-10% for slightly eroded cultivated land [7]

Many topical issues of agro ecological such as improve the productivity of the soil, protection from erosion, restoration of the eroded soil and the perfectibility of tillage technology are essential for decide forthwith.

The forest area of this region is reduced by 14175 thousand hectare in 2002 year while it is continuously decreased by 13301 and 12101 thousand hectare for 2008 and 2012 years, respectively. Totally 2074 thousand hectare area of forest has been destroyed in the last two decades. It is obvious that main reason of forest area decline is either usage of forest clear cutting or effects of forest insect, diseases, and the exploitation of mining [10]. Only for the last 4 years (2009-2012), accounting for burning 501 thousand hectare, 603 thousand hectare forest affected by insects and diseases, 113 thousand hectare of forest area disturbed by the exploitation of mining of the total forest fund area. The area of burned forest is more expanded in the Selenge, Bulgan, Khuvsgul and Arkhangai province while forest influenced by insect and desease is very serious in Khuvsgul and Bulgan province. Many examples can be observed that conversion process from forest soils into steppe type soil is active due to sharp changes of soil heat and soil moisture in the un-forested area. Above mentioned the global warming is accelerating to this changing process of soil type.

The water, soil and air are polluted due to technogen, in particular the level of environment pollution to be reach critical amount near to mining industry and settled area of soums, aimags, Erdenet, Darkhan and Ulaanbaatar city. Sequent studies found that the indicators of soil pollution such as arsenic, zinc, nickel, chrome and lead are over polluted in the many points around Ulaanbaatar city. Therefore bacteria and ammonian pollution created by deposition of domestic waste are infiltrated to soil while it is possible to effect on human health. Facts that river water, soil located near to gold mining are polluted by cyanite and mercury [11].

Mining sector is important for Mongolian economic development but it has been one of main factor for land degradation. During the building construction, earth-road of car, the building materials factory, the exploitation of the mining, there has changing landscape scenes,. Restoring degraded ecosystem is almost being left behind. As of 2002, in the basin of the Selenge River the mineral exploration license and the exploiting special license have about 1000 and 600, respectively. As follow this license, area for mineral exploration is about 7 M hectares whereas exploitation permitted area is approximately 150 thousand hectare. Most of explotation license areas are gold, building materials, coal mining. Degraded land in Selenge province is very high due to the mining operations and it is about 2735.0 hectares. In 2002 only, 1900 hestares area in Selenge river basin have been destroyed and abandoned without rehabilazation due to mining industry. Due to mining industry, mineral exploration natural ecosystem such as soil, water, and vegetation is degraded and soil, water and air pollution appears.

Conclusion

According to all of these, it is necessary to keep control of all sources of pollution and to be carry precaution from potential threat for ensures human health and environment. Further, the ecological safety research in the region of industry and settlement is necessary that it is need to choose right approach to clear up complex challenges for that one hand rehabilitation, conservation, and proper use of natural resources on the other hand human life, people to work and great decent conditions for human.

Reference

[1]- Статистикийн эмхэтгэл 2014: Улаанбаатар: Монгол улсын үндэсний статистикийн газар. Publication in Mongolian language. [Statistical year book 2014: Ulaanbaatar: National statistical office of Mongolia]

[2]- Мягмарцэрэн Д, Намхай А (2014): Монгол улсын тусгай хамгаалалттай газар нутаг. Улаанбаатар: Publication in Mongolian language [Myagmartseren D, Namkhai A (2014): Protected areas of Mongolia book. Ulaanbaatar;]

[3]- Байгаль нуурын ай савын экологийн атлас (2014): Улаанбаатар: Ш.Цэгмидын нэрэмжит Газарзүйн хүрээлэн, МУ-ын ШУА болон ОХУ-ын ШУА-ийн сибирийн салбарын В.Б.Сочавын нэрэмжит Газарзүйн хүрээлэн. Publication in Russian language [Ecological atlas of Basins of Baikal Lake, 2014 (2014): Ulaanbaatar: Mongolian Academy of Sciences Sh. Tsegmid Institute of Geography and Russian Academy of Sciences V.B.Sochav Institute of Geography]

[4]- Жигж С, (1975): Хотгор гүдгэрийн үндсэн хэв шинж. Улаанбаатар: Ш.Цэгмидын нэрэмжит Газарзүйн хүрээлэн, МУ-ын ШУА. Publication in Mongolian language. [Jigj.S, (1975) the main feature Mongolian relief book. Ulaanbaatar: Mongolian Academy of Sciences Sh. Tsegmid Institute of Geography]

[5]- Энхтайван Д, (2004): Сэлэнгэ голын сав нутгийн хотгор гүдгэрийн морфогенетик хэв шинж, сав газрыг хамгаалах шинжлэх ухааны үндэслэл. Улаан-Үүд: Байгалийн нөөцийн менежментийн олон талт арга. Publication in Mongolian language. [Enkhtaivan D, (2004): The Morphogenetic types of reliefs in Selenge river basin and their peculiarities Science for watershed conservation. Ulan-Ude: Multidisciplinary approaches for natural   resource management, Volume 1, 2]

[6]- Мөнхдулам О, (2013): Монгол орны төв хэсгийн физик газарзүйн тодорхойлолт сэдэвт судалгааны ажлын хүрээнд хийгдсэн Монгол орны төв хэсгийн ландшафт-газрын бүрхэвчийн ангилал, түүний өөрчлөлт дэд сэдэвт эрдэм шинжилгээний тайлан. Улаанбаатар: Ш.Цэгмидын нэрэмжит Газарзүйн хүрээлэн, МУ-ын ШУА. Publication in Mongolian language. [Munkhdulam O, (2013): Research reports of "The Landscape and land cover classification and its change, in the central part of Mongolia" covered by the report on "Determination of Physical Geography in the Center zone of Mongolia" Mongolian Academy of Sciences Sh. Tsegmid Institute of Geography]

[7]- Batkhishig O (Ed), (2008): Environmental changes of North East. Ulaanbaatar: Proceeding of the seventh Mongolia-Korea joint seminar on 2008: Institute of Geography, MAS, pp 1-2. Publication in English language

[8]- Доржготов Д, (2003): Монгол орны хөрс. Улаанбаатар: Шинжлэх ухааны академийн Газарзүйн хүрээлэн. Publication in Mongolian language. [Dorjgotov D, (2003): Soils of Mongolia book. Ulaanbaatar: Institute of Geography, MAS]

[9]- Монгол орны байгаль орчны төлөв байдлын 2011-2012 оны тайлан (2012): Улаанбаатар: Байгаль орчин, ногоон хөгжлийн яам. Publication in Mongolian language. [Report of status of the Environment of Mongolia, 2011-2012 (2012): Ulaanbaatar: Ministry of environment and green development of Mongolia]

[10]- Монгол орны байгаль орчны төлөв байдлын 2012-2013 оны тайлан (2013): Улаанбаатар: Байгаль орчин, ногоон хөгжлийн яам. Publication in Mongolian language. [Report of status of the Environment of Mongolia, 2012-2013 (2013): Ulaanbaatar: Ministry of environment and green development of Mongolia]

[11]- Томоохон хотуудын хөрсний бохирдол, экогеохимийн үнэлгээний тайлан (2013): Улаанбаатар: Шинжлэх ухааны академийн Газарзүйн хүрээлэн. Publication in Mongolian language. [Ulaanbaatar: The biggest cities assessment Eco-geochemical and soil pollution reports Mongolian Academy of Sciences Sh. Tsegmid Institute of Geography]

The natural risks caused by interactions between ecosystems of Selenga River Basin and the Central Asia

Petr D. Gunin, E.V. Danzhalova and Sergey N. Bazha

A.N. Severtsov Institute of Ecology and Evolution, Russian Academy of Sciences, 119071, 33 Leninsky prospekt, Moscow, Russia

email: pdgunin@mail.ru; sbazha@inbox.ru

Annotation

This case study describes the results of the spatial distribution of degradation processes in Selenga River basin and the levels of their impacts on the biota on the example of the Southern part of Lake Baikal territory within Mongolia. Inventory of ecosystems and their state assessment were executed on the basis of complex landscape-ecological mapping on the scale of 1: 500 000. On the basis for studying of changes in plant communities we had classified the most negative degradation processes dividing them on 5 main types. The sandisation and wind erosion provokes an invasion of *Caragana Bungei* which actively penetrates to the larch forest and steppe ecosystems of the basin. As a result, a projective cover and an aboveground phytomass of caragana species has increased in 2 – 2,5 times over the last 10 years. On the southern border of Selenga River basin an intensive process of *Ephedra sinica* distribution was registered. Being more aggressive on the vital strategy these species easily force out the native vegetation from ecosystems of dry steppes.

Introduction

Baikal Basin area occupying 540 000 km² situated on the junction of three world's largest watershed: Arctic Ocean, Pacific Ocean, and The Central-Asian Internal Drainage Basin. About 52 percent of the Baikal watershed area and over two thirds of the main artery of the Selenga River is in the central and northern parts of Mongolia (Figure 1). This region of Mongolia is one it's most densely populated and industrialized, and this creates conflicts between economic development of the region and enhanced conservation of resources. Mitigation of the conflicts requires better understanding of ecological conditions in the Southern (Mongolian) part of the Lake Baikal basin

The main objectives of our study are: i) to conduct an inventory of the ecosystems, and to assess their level of modification; ii) to analyze negative processes affecting Selenga River basin's ecosystems and components such as vegetation, soil, topographical relief, etc for classifying them according to ecological affect; and iii) to assess negative risks in the southern (Mongolian) part of the Basin to help mitigate environmental degradation; .

Figure 1. The Selenga River basin on the junctions of three world's largest watersheds and the research polygons (A, B).

Methodologies and Approaches

The investigation of the modification of the basin's environment involved landscape or ecological principles are used when the goal of assessment is to try to differentiate the biosphere into large scale mapping units (e.g., ecosystem types). The studies conducted by the Russian-Mongolian Complex Biological Expedition, the Russian Academy of Sciences, and the Mongolian Academy of Sciences, in the territory of Central Asia have been effective where the soil and plant cover has been well studied and when specialized and integrated maps of the regions have been available. As a basic unit we used the level of mesoecosystems. For more convenient presentation of the results we have combined 49 types of terrestrial ecosystems in 15 groups of ecosystems based on their zonal and altitudinal position.

We have analyzed ecosystems that have become modified at five levels due to anthropogenic impact. The techniques used are based on quantitative indices of changes in particular ecosystem components (vegetation, soil cover, relief) and their differentiation in terms of anthropogenic modification into major levels: absent, slight, moderate, heavy, very heavy, and transition between these categories (Gunin, Bazha, 2003).

The factors can be divided into two groups: 1) those leading to destruction of natural vegetation, disruption of natural links, and initial ecotype formation; and 2) transformation (sharp or occurring in the course of invasion or/and successional replacement) of indigenous vegetation and the ecotype which largely are recovered after the anthropogenic disturbances are discontinued. In terms of the types of economic transformation, the basin's ecosystems were classified into: 1) anthropogenic, 2) anthropogenic-natural, and 3) natural (Tateishi, Gunin, 2006).

Ecosystem diversity and their Anthropogenic disturbance

In total, the Selenga River basin encompasses ¾ of the ecosystem diversity of the Palaearctic part of the Asian continent (Table 1). This diversity is the combined result of: i) the location of the basin at the junction of three biogeographically distinct regions, namely the Central Asian, Eastern Asian, and European-Siberian region; ii) climatic variation across latitudes and longitudes; iii) the existence of multiple mountain ranges with large differences in altitude. Main ecosystem types in the basin include mountain tundra's, taigas, steppes, as well as river and lacustrine valleys.

Main Ecological Groups		Ecosystem types	Rate (%)	Area, sq.km.
Automor-phic and semi-hy-dromor-phic natural	I	Nival-golets	3,55	10346,35
	II	Mountain-tundra-grassland	5,30	15419,64
	III	Sub-golets	3,58	10432,15
	IV	North-taiga	0,14	420,50
	V	Mid-taiga	12,48	36333,85
	VI	South-taiga	9,90	28821,54
	VII	Mountain-meadow-steppe	6,79	19795,14
	VIII	Forest-steppe	6,92	20146,32
	IX	Moderate dry steppe	7,63	22215,43
	X	Dry steppe	27,23	79279,54
Hydro-mor-phic natural	XI	High-mountain and forest	4,33	12597,11
	XII	Forest-steppe	3,07	8933,60
	XIII	Steppe	5,74	16717,77
Anthropo-ge-nic	XIV	Arable and fallow land	2,17	6323,28
	XV	Urbanized and abandoned	1,18	3432,51
Total			100	291198,0

Table 1. Diversity of terrestrial ecosystem groups in the Selenga River basin (Mongolian part)

In general, landscapes at high altitudes of over 1800 m asl are dominated by mountain tundra. Alpine forests and Siberian cedar forests are in the lower regions of these high altitudes. Mid-elevation landscapes between 1200 – 1800 m asl are characterised by coniferous forests and cedar groves. Low-mountain landscapes between 600 – 1200 m asl are dominated mostly by conifers, cedar and larch, pines, and mixed taiga habitats. The low plains are characterized by forest, meadow, steppe and marsh landscapes.

The structures of terrestrial ecosystem cover of the Southern part of Selenga River basin include 14242 areas (mapping circuits) and reflected in 49 ecosystem types described at the level of zonal and intrazonal mesoecosystems, which account for 88,4 percent of zonal ecosystems in the basin's area. Intrazonal ecosystems including hydromorphic account for 10.6 percent; and lacustrine ecosystems account for no more than 1.0 percent. The dominance of mountain-forest and plain-steppe ecosystems is indicative that the Mongolian part of the Baikal basin should be classified as a region with a predominant distribution of mountain-forest-steppe ecosystems, which confirms its ecotone nature. The main ecosystem groups were further sub-divided according to their level of anthropogenic disturbances (Table 2).

Ecosystems group	Area, sq. km	Area by the level of disturbance, sq. km				
		I	II	III	IV	V
I	10346,35	2112,80	1928,11	5579,38	726,07	0,00
II	15419,64	1820,38	3109,62	9268,77	1220,87	0,00
III	10432,15	1782,32	3291,27	4927,51	431,05	0,00
IV	420,50	217,70	155,29	31,85	15,65	0,00
V	36333,85	3074,81	17029,00	14758,06	1471,99	0,00
VI	28821,54	7148,81	10985,33	8955,57	1718,47	13,35
VII	19795,14	632,22	2692,21	8690,91	7769,89	0,00
VIII	20146,32	941,14	7853,15	8389,08	2705,16	250,99
IX	22215,43	201,86	774,41	9649,33	11578,71	11,12
X	79279,54	419,93	5246,73	31200,66	42112,38	299,85
XI	12597,11	651,46	361,09	7289,76	4292,76	2,05
XII	8933,60	53,81	132,94	3918,49	4828,35	0,00
XIII	16717,77	26,59	991,94	6878,29	8820,96	0,00
XIV	6323,28	0,00	26,66	1421,36	4875,26	0,00
XV	3432,51	0,00	120,47	1403,90	1897,85	10,30
Total	291198,0	19 083,8	54 698,2	122 362,9	94 465,4	587,65

Table 2. Area of main ecosystem groups divided by the level of disturbance

The eco-biological assessment of the state of ecosystems led us to conclude that the detrimental anthropogenic impacts on the structure and functioning of the ecosystems are manifested at the physiognomic level when the modification is at the IIIrd level. Under the Vth level of modification the ecosystems are classified as completely transformed anthropogenic. Analysis of modification as determined by various anthropogenic factors reveals that overgrazing has to date heavily and very heavily modified 13.2 percent of ecosystems' area, and rangeland ecosystems that have been moderately modified account for 54.6 percent of their total area. Felling and fires have modified forestry ecosystems heavily and very heavily in over 20 percent of the forest-covered territory. Virtually all of the agroecosystems of the region have been anthropogenically modified to the IIIrd to Vth degree.

Based on our assessment of degradation processes responsible for heavy and very heavy anthropogenic modification of the natural environment we have distinguished five groups of hazardous degradation processes: 1) rangeland overgrowth with shrubs, 2) deforestation of forest-steppe ecosystems, 3) desertification of ecosystems on light soils, 4) depletion of ecosystems with sand soils, and 5) narcotization of agrocenoses in modified ecosystems. We also have differentiated the impacts on the basis of their significance to ecosystems, the economic systems, or social systems, respectively (Gunin, Bazha, 2003).

In terms of economy, degradation processes include deforestation of forest-steppe ecosystems and the processes of rangeland depletion in hydromorphic landscapes. A deforestation of forest massifs is most intensive in the forest-steppe zone where forest ecosystems are "insular" because of their exposure to maximum phytocenotic pressure on the part of steppe vegetation. Thus, when a tree stand is replaced by a shrub or grass stand the impact often is significant because of a reduction not only of biological diversity, but also because of a reduction in forest carrying capacity.

Study of interaction between the Central Asian and Selenga River basin ecosystems

Degradation processes that are significant to social-economic systems stem from the invasion of synanthropous and anthropofilic species of plants particularly those containing alkaloids. Despite the heterogeneous nature of the processes they have a common feature of affecting human health.

As it has mentioned above, the extent of joint border between the Selenga River basin and the Central Asia's area is about 1450 km, or over 30 percent of total length of Basin's border. Character and consequences of ecosystem interactions between Selenga River basin and The Central Asian Region steel needs to be investigated more comprehensively. The influence of external factors on two transboundary research polygons: Tosontsengel – at the west side, and Undjul – on the southern border of Selenga River basin was studied (Figure 1).

Investigated sites represent the ecotones which have been chosen by us on the western and southern periphery of Selenga River basin in a zone of interaction with desert-steppe and desert landscapes of the Central Asia. The most suitable to studying of such ecosystems is the method of ecological profiling allowing the closest captures a difficult structure of a contact zone (Bazha, et al., 2013). Desertification processes of the forest-steppe ecosystems caused by penetration of desert-steppe bush *Caragana bungei* Ledeb. to native vegetation on the example of Tosontsengel research polygon, Western side of Selenga River basin were studied in 2004 and 2014. Investigated communities are located in a valley of Ider River, and they presenting the east part of *Caragana bungei* (56-1, 56-2, 56-3, and 56-4) (table 3).

Desertification processes characterised by invasion to the dry-steppe plant communities of deserted-steppe evergreen low shrub *Ephedra sinica* Staph. were studied in 2005 – 2010 on Unjul research polygon at southern border of Selenga River basin on gently ridged plains in Bayan-Undzhul Soum (Central Aimag of Mongolia). In the investigated plant communities (14, 15, 19, and 26) the participation rate of *Ephedra sinica* was various - from absolute domination to absence of species (Table 3).

Description Index	Plant Associations	Progective cover of Caragana bungei	Progective cover	Coordinates	Elevation, M, exposure, slope
56-1	Larch forest with sedge - motley grass - caragana	12,0	-	N48° 39' 1,7" E98° 56' 1,4"	1858, SE, 30°
56-2	Mountain steppe with petrophyte motley grass – sedge - caragana	26,3	-	N48° 39' 2,1" E98° 56' 8,1"	1875, NE, 40°
56-3	Meadow-steppe with larch and motley grass - sedge - caragana	19,0	-	N48° 39' 2,3" E98° 56' 10,2"	1856, NE, 25°
56-4	Mountain steppe with petrophyte sedge - motley grass –- cereal - caragana	16,1	-	N48° 39' 13,5" E98° 56' 07,3"	1870, SE, 27°
14	Cereal – ephedra steppe with caragana and synusia of annual species	-	10,0	N46° 58' 49,0" E105° 55' 48,8"	1323 Plain
15	Steppe with caragana - ephedra and synusia of annual species	-	10,0	N46° 58'52,4" E105° 55' 50,0"	1321 Plain
19	Steppe with onion – summer cypress - feather grass and synusia of annual species	-	-	N46° 57'59,6" E105° 34' 27,0"	1339 Plain
26	Steppe with ephedra - caragana and synusia of annual species	-	25,0	N47° 09'10,8" E105° 23' 56,0"	1166 Plain

Table 3. Participation of Caragana bungei Ledeb. and Ephedra sinica Staph. in vegetation composition of Unjul and Tosontsengel research polygons

Results and discussion

On the western periphery of Selenga River basin *Caragana bungei* defines a structure of the transformed steppe communities by the bottom of flat parts along the slopes of hills and the leveled surfaces adjoined to them with elevation of 1400-1580 m asl. As a result, rapid invasions of *Caragana bungei* into grasslands the pasture turns in caragana thickets with a projective covering of bush from 20 percent up to 50 percent. On the sandy massifs generated as a result of aeolian movement, *Caragana bungei* penetrates into communities of petrophyte mountain steppe, as well as into the larch forest. A considerable abundance of Caragana is characteristic for mountain steppes. In Mountain steppe with petrophyte motley grass – sedge - caragana community (56-2) 29 bushes of Caragana per 100 M^2 are counted, and their projective covering has made more than 26 percent. Participation the other kind of bushes is rather insignificant. In the Mountain steppe with petrophyte sedge - motley grass - cereal plant community (56-4) the greatest participation of *Caragana bungei* was fixed. Its share here is 73 percent from the total projective covering (Table 3).

For penetration of *Caragana bungei* into larch forest the repositioned sand massifs on steep (20 – 30 °) slopes of hills of a northeast exposition serve as natural ecological corridors. Thus, the wood floor of larch forest (56-1) is represented by middle-aged larch *(Larix sibirica)*, 10 – 12 m height and with rare undergrowth. Sparse shrub layer of single individual cotoneaster *(Cotoneaster melanocarpa)* and meadowsweet *(Spiraea media)* did not become a barrier for Caragana penetration (12 species per 100 M^2). The projective covering of a grass layer has 28 percent, with the dominance of *Erysimum hieracifolium* and *Carex pediformis*. In communities with sparse tree layer Caragana is more plentiful. In the meadow-steppe with sparse larch and motley grass - sedge plant community (56-3) a projective cover of *Caragana bungei* has made 19 percent.

Thus, a peculiar ecological corridors generated as a result of aeolian transport, as well as the absence of competitive species in shrub cover are the main contributors for the spread of *Caragana bungei* in the community of hemi boreal larch forests in the taiga-meadow-steppe mountain zone.

Studies of plant communities in arid ecosystems of Bayan-Undzhul soum showed widespread adoption of *Ephedra sinica*, both in lowland ecosystems and rocky slopes of low hills and in the ecosystems of plains. Mapping of steppe ecosystems in Bayan-Undzhul soum showed that ephedra is a part of the communities currently involved in more than 1/3 of the research polygon's area. (Gunin et al., 2012).

In severely degraded communities (14, 15) with a low abundance of grasses *(Agropyron cristatum, Cleistogenes squarrosa, Koeleria criststa, Stipa krylovii)* Ephedra has the leading role. Its projective cover was 10 percent, while the number of partial bushes per 1 m^2 is 13,5 in average. Phytocenotic role of ephedra increased in transformed communities with complete loss of herbage grasses, as it was revealed in the Steppe - caragana with synusia of annuals plant community (26). Lack of *Ephedra sinica* observed in the Steppe with onion – summer cypress - feather grass and synusia of annuals plant community (19). It is in relatively better condition compared with 1970 – 1980, and retained virtually all of their species composition (Dry steppes of MPR, 1984).

According to our observations in different parts of Mongolia, ephedra is not reliable due to the nature of the relief and soil composition. It introduced and grows well on various soil types, dedicated to a variety of types of relief (Gunin et al., 2012).

Colony forming ephedra, which may be relatively isolated from each other. Such a mosaic distribution of ephedra in Mongolia, without rigid binding to specific ecotopes, suggests that one of the important reasons for its settlement is zoochory, i.e. spread ephedra using phytophagous mammals and birds. In particular, in the Mongolian steppes, its seeds can spread hoofed animals, eating a juicy ephedra fruits.

Conclusion

The results of soil and geo-botanical studies conducted on the border between the Selenga River basin and the Central Asian inland runoff made possible determining the cross-border cooperation places in the vegetation cover.

The nature of such interactions were identified as invasive, primarily associated with the penetration of the Central Asian desert-steppe species into the Selenga River basin ecosystems: in forb-grass steppe and larch forest - *Caragana bungei*, and into dry steppe - *Ephedra sinica*.

It should be noted that the current invasive succession in both parts of the Basin are differ by as the dominant species taking an active part in invasions, and in their territorial scope and landscape confinement. In the first case, the clear confinement of *Caragana bungei* communities to habitats with sandy sediments deposited aeolian flows from the Great Lakes basin in Central Asia was found. Inclusion of the seeds of this species in the wind-sand flow provides long-range transport of indigenous communities, and stimulates penetration of caragana into the new types of ecosystems which are not typical places for this kind of species.

In the second case (Unjul research polygon), there is a formation of monodominant plant communities of *Ephedra sinica* or its significant participation. To date, the spatial distribution of ephedra communities is patchy, while that is not associated with certain types of landscape, and a wide ecological range of this species suggests a progressive direction of this type of succession.

Modern resettlement of these species contributes to arid climate, changing soil conditions and pasture digression. Taken together, these factors favor the development of species historically adapted to desert-steppe habitats.

Ecological and biological features of these two species widespread in desert landscapes make it possible to diagnose the above mentioned processes as biological desertification.

Acknowledgements

The study was started under the framework of the Joint Russian-Mongolian Complex Biological Expedition, Russian Academy of Sciences and Mongolian Academy of Sciences. Data analysis and the results preparation were conducted with the financial support of Russian Geographical Society, Grant 04/2014/RGS_RFBR.

The authors extend their sincere thanks to Drs.: I.M. Miklyaeva, Sh. Bayasgalan, T.I. Kazantseva, Yu.I. Drobyshev for constant support and participation in the fieldwork for studying the Selenga River basin ecosystems.

References

Gunin PD, Bazha SN (2003): Ecological assessment of degradation processes in the Northern part of Baikal Basin. In: Conserving Biodiversity in Arid Ecosystems. KAP. Chapter 12. 157–177.

Бажа СН, Востокова ЕА, Гунин ПД, Дугаржав Ч., Данжалова ЕВ, Воробьев КА, Прищепа АВ, Петухов ИА. (2013): Геоинформационное картографирование наземных экосистем бассейна Селенги на примере модельных участков. Методические рекомендации. М.: Росельхозакадемия, 109 с. Publication in Russian language. [Bazha SN, Vostokova EA, Gunin PD, Dugarzhav Ch Danzhalova EV, KA Vorobiev, Prischepa AV Petukhov IA (2013): GIS mapping of terrestrial ecosystems of the Selenga Basin for model sites. Methodical recommendations. M.: Roselhozakademiya. 109 p.]

Gunin PD, Bazha SN, Danzhalova EV, DmitrievIA, Drobyshev Yu.I. Kazantseva TI, Ariunbold E, Battseren C, Jargalsaikhan L. (2012): Expansion of Ephedra sinica Stapf. in the arid steppe ecosystems of Eastern and Central Mongolia. Arid Ecosystems. SP MAIK Nauka/Interperiodica.Vol. 2. Issue 1. 18-33.

Лавренко ЕМ (Отв. ред). (1984): Сухие степи Монгольской Народной Республики: природные условия (сомон Унжул). Л.: Наука. Ч. 1. 167 с. Publication in Russian language. [Lavrenko EM (E. Ed.) (1984): Dry steppes of the Mongolian People's Republic: natural conditions (Unzhul soum). Leningrad: Nauka. Vol. 1. 167 P.]

Tateishi R, Gunin P (E. Eds.) (2006): Present-day ecosystems of the Selenga River Basin and factors of their destabilization. CEReS, Chiba University press. Japan. 139 p.

Is the Endemic Fauna of Lake Baikal Affected by Global Change?

Till Luckenbach[1], Daria Bedulina[2], Maxim Timofeyev[2]

[1] Department of Bioanalytical Ecotoxicology, UFZ—Helmholtz Centre for Environmental Research, Permoserstr. 15, 04318 Leipzig, Germany

[2] Institute of Biology at Irkutsk State University, Lenina Str. 3, 664003 Irkutsk, Russia

Abstract

Lake Baikal, the largest freshwater lake in the world and UNESCO world heritage site, is inhabited by an exceptionally species-rich, largely endemic fauna which is distinct from faunas from other freshwaters in the Palearctic. With regard to species-richness, extremely high abundance of individuals and high total biomass amphipods (Amphipoda, Crustacea) constitute a major, ecologically highly relevant animal taxon of Lake Baikal. The "immiscibility barrier" of faunas, i.e., the separation of amphipod faunas from Lake Baikal and other Palearctic waters, is believed to be related with the high degree of adaptation of Lake Baikal species to the specific environmental conditions of their habitats; under these conditions they outcompete potentially invasive non-Baikal species that are unable to establish stable populations in Lake Baikal. It is a question of great interest whether the current global change related, massive alterations of temperature and chemical conditions in Lake Baikal will favor potential invasive species. Our molecular and physiological studies with representative endemic Eulimnogammarus species show that these species are indeed equipped to deal with stress from unfavorable environmental conditions and in comparison to a potentially invasive amphipod may not necessarily be more sensitive. However, considerable differences in stress responses among the different species suggest that species shifts in the Lake Baikal ecosystem may occur when environmental conditions continue to change.

The exceptionally unique ecosystem Lake Baikal

Lake Baikal, located in an intracontinental rift zone in the central region of southern Siberia, is the world's oldest (25–30 million years), by volume largest (23,000 km3) and deepest (1,642 m) lake, containing about 20% of the world's liquid freshwater (equivalent to all North American Great Lakes combined) (Rusinek et al., 2012a). As unique ecosystem with exceptionally high degrees of biodiversity and endemism it was designated as a UNESCO World Heritage Site in 1996 (http://whc.unesco.org/en/list/754). So far, 2,595 animal species from Lake Baikal have been identified or described, of which 80% are endemics (Rusinek et al., 2012b; Timoshkin, 2001). This high degree of endemism reflects the long evolutionary history of Lake Baikal in isolation from other freshwater bodies (Timofeyev, 2010).

The fauna of Lake Baikal is represented by two genetically and ecologically different complexes – the Euro-Siberian (Palearctic) fauna inhabiting so called "sors", which are isolated waters at the shores of Lake Baikal that have connection to the lake, and the "Baikalian" fauna of the open lake (Kozhova and Izmest'eva, 1998). It is indeed a question of great interest why the "Baikalian" fauna remains majorly distinct from the fauna of sors and adjacent ponds and lakes with species compositions that are commonly found in fresh water systems across northern Eurasia (see Figure 1).

A major factor contributing to the formation of a faunistic "immiscibility barrier" between Lake Baikal and other fresh waters can be seen in the, unique abiotic conditions of Lake Baikal (Mazepova, 1990). The water is classified as ultra-oligotrophic with comparatively very low levels of ions and dissolved organic carbon; oxygen levels are permanently exceptionally high throughout the water column with concentrations close to saturation; and although the water temperature can rise to 20-22°C close to shore in protected shallow bays in summer (Khozov, 1963) it is overall constantly low with 6°C in average (Falkner et al., 1991; Weiss et al., 1991; Yoshioka et al., 2002). The abiotic conditions of Lake Baikal thus show fundamental differences to all other Siberian freshwater systems that are characterized by strong seasonal fluctuations of environmental parameters temperature, oxygen content, osmotic conditions and contents of natural organic compounds. It therefore seems obvious that differences between Lake Baikal and other lakes regarding abiotic conditions may require differences in physiological performance ranges as adaptations of species of the respective faunas to respective conditions. It is assumed that invasions of non-Baikal species into Lake Baikal and formation of stable populations in the lake are precluded by superior performance of the highly adapted species of the endemic fauna of Lake Baikal under these specific conditions which therefore out-compete other, potentially invasive species thus keeping ubiquitous northern Eurasian and Lake Baikal faunas distinct (Timofeyev, 2010).

Figure 1. Photograph of a site close to the village Onguryony in the north-west of Lake Baikal. The picture illustrates the "immiscibility barrier" of Lake Baikal and Palearctic freshwater faunas. A typical Lake Baikal shore habitat is in close vicinity and only separated by a narrow stretch of land from a freshwater habitat with a completely different character inhabited by a fauna common in the Palearctic.

Amphipods – dominant benthic organisms in Lake Baikal

Amphipods (Amphipoda, Crustacea) are highly abundant macro-invertebrates that substantially contribute to the overall biomass in Lake Baikal and constitute key components of the benthos from the littoral to abyssal zones. The number of amphipod species in Lake Baikal is exceptionally high and the lake is worldwide one of the hotspots of amphipod species diversity. From animal taxa inhabiting Lake Baikal the Amphipoda taxon has the highest documented diversity and endemicity. All of the so far classified 354 amphipod species and subspecies are endemic to Lake Baikal (Bedulina et al., 2014) and the species number amounts to approximately 20 % of the number of so far described amphipod species from fresh or inland waters worldwide (Väinölä et al., 2008).

The complex taxonomy of amphipods from Lake Baikal is still a matter of debate. Few species, mainly introduced by human activity, such as Gmelinoides fasciatus, Eulimnogammarus cyaneus, E. viridis, and Micruropus wohlii, were found to spread to waters outside of Lake Baikal (Berezina, 2007; Gladyshev and Moskvicheva, 2002; Takhteev, 2000). However, overall, the endemic amphipod fauna is confined to the

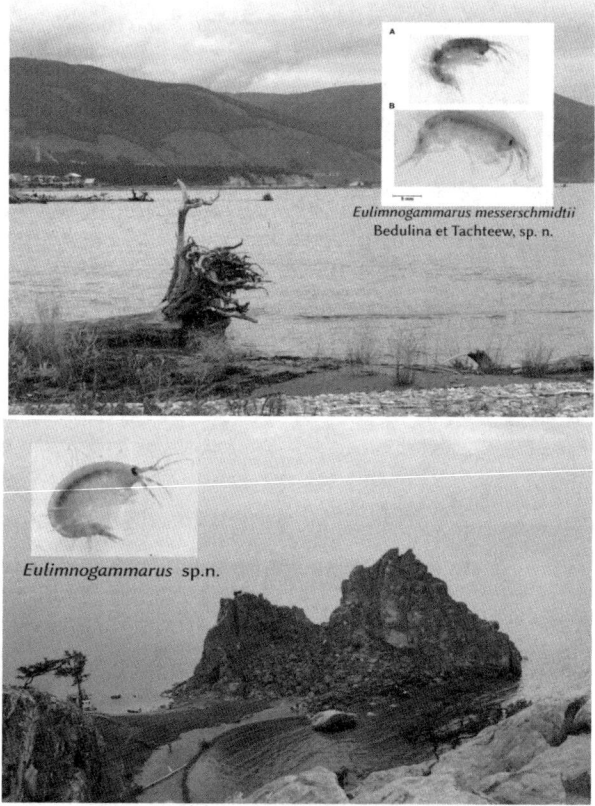

Eulimnogammarus messerschmidtii
Bedulina et Tachteew, sp. n.

Eulimnogammarus sp.n.

Figure 2. Photographs of recently discovered amphipod species endemic to Lake Baikal and of the shore lines where they were found.
Above: *Eulimnogammarus messerschmidtii* Bedulina et Tachteew, sp. n. (blue and red morphs – A, B) and the shore of northern Lake Baikal close to the city of Severobaikalsk.
Below: A yet undescribed species (*Eulimnogammarus* sp. n.) which was found close to the Shaman rock on Olkhon island.

waters of Lake Baikal and ponds and lakes in its vicinity are inhabited by a clearly distinct fauna with Gammarus lacustris as typically predominant amphipod species (Kozhova and Izmest'eva, 1998). Although amphipods from Lake Baikal have intensively been studied over decades there is still a large number of species to discover. Thus, it is assumed that only half of the species making up the endemic fauna of Lake Baikal has been described (Timoshkin, 1999). It is not just deep-water amphipod species that await discovery, but also new littoral species can be found at shores that are relatively well accessible and frequented by humans. Just recently, we published the description of a littoral amphipod species, Eulimnogammarus messerschmidtii that we found on a field trip to northern Lake Baikal close to the city of Severobaikalsk (Bedulina et al., 2014). Another undescribed littoral amphipod species that we found on the Olkhon island at a beach close to the Shaman rock, landmark of Lake Baikal and tourist attraction, still needs to be named (Figure 2). So far, the main focus of studies of amphipods from Lake Baikal has been on taxonomy and ecology and not so much on physiological adaptations to environmental conditions and their molecular basis.

We recently studied two major components of cellular stress response mechanisms in amphipods, so called chaperones or heat shock proteins (hsp) stabilizing the structure of other proteins and cellular transporter proteins from the ABC (ATP binding cassette) transporter protein superfamily that keep out toxic chemicals from cells (Bedulina et al., 2013; Pavlichenko et al., 2015). An important outcome of these studies is that amphipods endemic to Lake Baikal do possess these cellular stress response mechanisms, i.e., these species are equipped to deal with stress caused by adverse environmental conditions, such as temperature or chemical stress.

The identification of genetic sequences in amphipods is still challenging as there is a lack of genomic data for this animal taxon which could be used as reference. A recent first genome next generation sequencing (NGS) technology based sequencing campaign of the genome of Eulimnogammarus verrucosus (Rivarola-Duarte et al., 2014), an amphipod endemic to Lake Baikal, can in this respect be seen as pioneering and the obtained data will be of great value for future studies on Lake Baikal amphipods, but may also serve as reference for studies on amphipods in general. The sequenced genome of E. verrucosus appears to be unique in many regards. With a size of about ten Gb (giga bases) it turned out to be surprisingly large, about three times larger than the human genome. The large size cannot be associated with genome duplication as no paralogs of highly conserved genes, so called hox genes, were found, however, the genome appears to contain a high proportion of non-coding DNA, i.e. DNA not coding for any proteins. A comparison of genomes from E. verrucosus and the water flea Daphnia pulex, the closest relative with a well-developed genomic resource, showed only very little sequence similarities indicating that the genomes are quite distinct (Rivarola-Duarte et al., 2014).

Lake Baikal and global change

As a consequence of the drastic alterations of global climate with world-wide increasing mean temperatures and of the global chemosphere resulting from massive releases of man-made chemicals into the environment the abiotic conditions of Lake Baikal have dramatically changed in the last decades and those changes are predicted to continue. Recent models predict a 1.8-5°C increase of the mean global temperature by the year 2100, with increased occurrences of temperature fluctuations and temperature extremes (IPCC, 2007). Simultaneously, the global chemosphere is currently heavily influenced by human activities. The number of chemicals in use by mankind amounts to 100.000, with 10.000 to 30.000 being of environmental concern (Hartung and Rovida, 2009).

Lake Baikal is in one of three areas in the world experiencing the most rapid climate change; the other two regions are the Antarctic Peninsula and northwestern North America (Clarke et al., 2007). All three areas are characterized by long, cold winters. At Lake Baikal, winter air temperatures reach −37°C to −40°C, and the lake freezes for four to five months each year; summer air temperatures soar briefly to 25°C to 30°C in this strongly continental climate (Kozhova and Izmest'eva, 1998).

Despite the enormous temperature buffering capacity of the large water body global warming has caused a 1.21°C increase in the average surface water temperature in the past 50 years (Hampton et al., 2008), a rate twice that of the global average, and the ice-free season has lengthened by 16.1 days between 1868 to 1995 (Magnuson et al., 2000). This is of particular importance for the Lake Baikal ecosystem since ice is arguably the single most important abiotic driver in this lake. The lake's dominant primary producers and its top predator, the Baikal seal, require ice for population growth. Whereas the spring phytoplankton bloom begins shortly after ice off in temperate-zone lakes the spring bloom occurs under the ice in Lake Baikal and ice is essential for initiating and sustaining this bloom (Moore et al., 2009). Long-term monitoring surveys demonstrated increases of more than 300% in average zooplankton, algae and cyanobacteria abundances since the 1940s as a consequence of increasing temperatures (Hampton et al., 2008). The increase of temperature in the Baikal region will remain substantial in the future. The projected median increase of annual air temperatures will be 4.3°C in 2080-2099 compared to 1980-1999, with 6°C for the median projected temperature increase for the winter months and 3°C for summer (Christensen et al., 2007).

In addition to temperature change industrial pollution and cultural eutrophication are of particular concern for Lake Baikal. The Irkutsk region, containing an industrial corridor with chemical plants and aging industries, lies within the lake's airshed and industrial chemicals including polychlorinated dibenzo-p-dioxins and dibenzofurans (PCDD/Fs) and polychlorinated biphenyls (PCBs) are carried into the southern basin of the lake by prevailing winds and bioconcentrate in fish, the Baikal seal and in humans (Mamontov et al., 2000). Likewise, perfluorochemicals (PFCs) originating from ongoing contamination from a local source bioconcentrate in Baikal seals (Ishibashi et al., 2008). Other sources of pollutants include the Trans-Siberian railroad, now transporting oil along the southern and eastern shores of the lake, the Baykalsk Pulp and Paper Mill (BPPM), a large deteriorating pulp mill on the southern lake shore, industrial sites in the area of the city Severobaikalsk and agricultural waste water in the Selenga river flowing into Lake Baikal. Eutrophication supposedly from elevated phosphate and nitrate levels in some shallow bays along the Baikal shore was assumed to result from increased tourism and ship traffic with some consequences for the ecosystems in these habitats, such as unusually high abundance of Spirogyra and other algae (Kravtsova et al., 2014; Timoshkin et al., 2014) and in Lisvyanka bay during the summer of 2014 occurrence of frequent deaths of the endemic freshwater sponges that by acting as filter feeders have an important function for water purification. The influx of chemicals and nutrients is anticipated to further increase as a result of thawing permafrost releasing chemicals from soil and increased run-off and erosion into the lake with elevated temperatures.

Due to the lake's distinct features, such as oligotrophy, cold waters, long residence time, a long pelagic food chain, the high seismicity of the region, and great endemism, its ecosystems may be particularly vulnerable to stress resulting from the current environmental changes (Moore et al., 2009).

Do changes of environmental conditions enable invasive species to enter Lake Baikal?

Freshwater ecosystems are particularly affected by both temperature changes and chemical pollution. Surface runoff and industrial and municipal waste water often end up in streams and lakes and there is a continuous inflow of chemicals into aqueous environments. Increased temperature in concert with chemical stressors result in increased overall stress for organisms (Lannig et al., 2008).

These alterations of environmental conditions are advantageous for species with a superior ability to cope with stress resulting from environmental changes, because they become able to outcompete species that are less stress tolerant. Climate and chemosphere changes favor ubiquitous "generalists" that outcompete indigenous/endemic faunas. Mass occurrences of invasive species, such as for instance shown for certain amphipod species, occur in anthropogenically heavily affected environments (Grabowski et al., 2007; Van den Brink et al., 1991).

It is therefore of great concern that the current environmental changes by global warming and increasing levels of anthropogenic pollutants will lead to a situation where the immiscibility barrier of faunas in Lake Baikal deteriorates. Conditions that have hitherto been the cause of competitive superiority of endogenous Baikal species over ubiquitous Eurasian species may no longer be in place. The circumstance that Baikal species are perfectly adapted to conditions within a narrow range of variation will no longer remain a competitive advantage since environmental conditions may be shifted beyond the limits of optimal performance.

Various cases have shown invasions of non-indigenous species in other ecosystems as a consequence of global climate change and anthropogenic pollution, with success of invasion depending on the degree of stress tolerance (Grigorovich et al., 2008; Stachowicz et al., 2002). Indeed, cases of invasions of species which could be related to environmental conditions, i.e. specifically increased temperature, have also been reported for Lake Baikal; for instance, invading freshwater snails have been found to completely replace the indigenous fauna in certain isolated, shallow bays where comparatively high water temperatures were reached, which was related to a higher degree of thermotolerance of the invasive species (Stift et al., 2004).

On this background, current climate change and pollution can be regarded as substantial threat to the unique ecosystem of Lake Baikal as they may considerably advance replacement of indigenous by invasive species, alterating the Baikal ecosystem. Apart from the tragedy for the world community in case of deterioration of this unique ecosystem, changes of the Lake Baikal ecosystem will also have severe consequences for the local society. For one, local economies dependent on the lake, such as fisheries, may be threatened. Secondly, water supply of local communities may be affected by impaired water quality.

Gammarus lacustris – a potential invasive amphipod species to Lake Baikal?

For addressing the question if changing environmental conditions will enable non-Baikal amphipod species to invade Lake Baikal we chose three species as representative models for experiments. Two of the species, Eulimnogammarus cyaneus and E. verrucosus, are endemic to Lake Baikal and highly abundant littoral species. E. cyaneus inhabits the narrow zone near the water edge and E.verrucosus typically occurs close to shore at water depths from 0 to 6 m. The third species, Gammarus lacustris, can be regarded as potentially invasive species. It is widely distributed in fresh waters of Northern Eurasia, including Siberia (Karaman and Pinkster, 1977). The species occurs in some isolated bays of Lake Baikal, but it is not a typical species of the lake´s fauna (Kozhova and Izmest'eva, 1998; Timoshkin, 2001).

In contrast to species from Lake Baikal G. lacustris may need to tolerate higher fluctuations of environmental conditions, such as temperature changes during the day, as the shallow waters inhabited by the species do not show the buffering capacity of the large water mass of Lake Baikal and due to their limited size do not offer possibilities to escape adverse conditions by migration to other areas. It may therefore be assumed that G. lacustris is better equipped with cellular and physiological adaptations to withstand extreme conditions than the Baikal amphipod species, which in the contrary may have to deal less with fluctuations of environmental parameters in their habitats.

However, our studies show that greater sensitivity to adverse environmental conditions cannot per se be assumed for all Lake Baikal endemics.

Thus, experimental data on temperature sensitivities of the three amphipod species show that indeed a Lake Baikal endemic, E. cyaneus, is least sensitive to temperature stress (Figure 3). At 25°C water temperature, which causes thermal stress in the amphipods, increases in mortalities with time of exposure were found in all three species. However, whereas E. verrucosus was most sensitive with almost 100 % mortality after 24 hrs, mortality was lowest in E. cyaneus and, compared to E. cyaneus, clearly higher in G. lacustris (Figure 3). Based on these findings it appears plausible that E. cyaneus, although endemic to Lake Baikal, nevertheless experiences temperature fluctuations with comparatively large amplitudes or extremes in its habitats. Indeed, temperatures of the shallow waters in the upper littoral, where E. cyaneus is found, can reach maxima above 20°C in the summer (Shimaraev et al., 1994). Further, E. cyaneus can be found in warm waters close to hot springs (own observation). Thus, a high degree of temperature tolerance appears to enable the species to survive in habitats that with regard to environmental conditions are relatively extreme enabling it to escape competition with species that are less temperature tolerant. E. verrucosus, however, as more temperature sensitive littoral species appears to respond to unfavourable temperature conditions by migrating to deeper colder waters as it disappears from the shallow littoral during warm summer periods. It is only found in the shallow littoral when the water is cooler, either due to colder climate or due to upwelling of cold water from the deep (Weinberg and Kamaltynov, 1998).

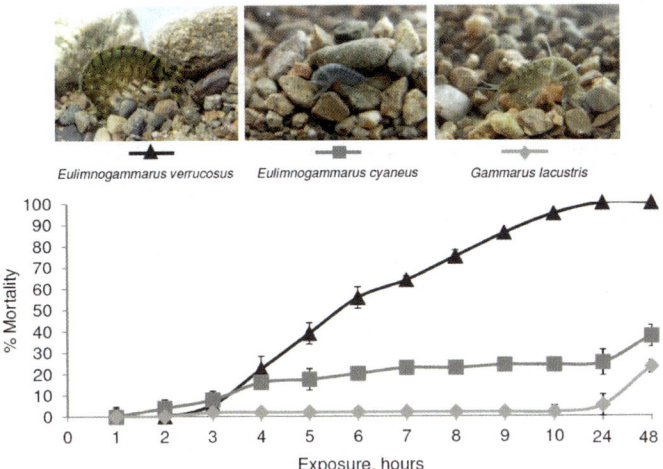

Figure 3. Percent mortalities of different amphipod species, *E. cyaneus*, *E. verrucosus* and *G. lacustris* in dependence of time of exposure to thermal stress (water temperature at 25°C). Upon lab adaptation of wild caught amphipods animals were kept in water tanks in aerated water maintained at 25°C and dead animals were counted and removed from the tank every hour. For details on origin and maintenance of animals and on the experimental set up refer to Bedulina et al. (2013) and Pavlichenko et al. (2014). Data from Bedulina et al. (2013) and Timofeyev (2010). Photographs of amphipods by Vasiliy Pavlichenko.

As outlined above the studied endemic amphipods possess the cellular machinery enabling the organisms to deal with adverse environmental conditions, such as thermal and chemical stress. However, there are species-specific differences in cellular stress response systems that can be seen as reason for differences between species in the degree of tolerance to environmental stress. Thus, the Hsp70 system mitigating destabilizing effects of high temperature on protein structure is more complex and more active in E. cyaneus than in E. verrucosus (Bedulina et al., 2013) and constitutive hsp70 transcript levels are generally higher in G. lacustris and E. cyaneus than in E. verrucosus, with particularly high constitutive HSP70 protein levels in E. cyaneus (Protopopova et al., 2014). These molecular data are thus in line with the different degrees of thermal tolerance indicated by the temperature-related mortalities in those species (Figure 3).

Conclusions

Our data do not indicate that non-Baikal amphipods are generally more stress-tolerant than Baikal endemics. Therefore, continuing environmental changes may not necessarily enable invasive species to occupy new habitats in Lake Baikal since species, such as E. cyaneus, may still be able to deal with changed conditions, especially higher maximum temperatures, and out-compete potential invaders. However, migration of endemic species, such as E. verrucosus, to deeper, cooler waters as a consequence of higher water temperatures may increase the competition pressure in these habitats leading to ecosystem shifts. Changes in the Baikal water chemistry, in particular eutrophication, may lead to dramatic impacts on the Lake Baikal ecosystem, as recently reported, and urgently needs to be addressed by water managers and scientists.

Acknowledgements

We want to thank our colleagues Lena Jakob, Lorena S. Rivarola-Duarte, Anton Gurkov, Denis Axenov-Gribanov, Vasiliy Pavlichenko, Christian Otto, Marina Protopopova, Zhanna Shatilina, Xenia Vereshagina, Hendrik Michael, Inna Sokolova, Irene Fernandez Casas, Wibke Busch, Michael Ginzburg, Hans-Otto Pörtner, Peter Stadler, Rolf Altenburger, Stephan Schreiber, Jörg Hackermüller, Magnus Lucassen, Franz-Josef Sartoris and Michael Evgen'ev, Olga Zatsepina, Vadim V. Takhteev who are or were involved in the activities of our Helmholtz-Russia joint research group LaBeglo ("Lake Baikal and biological effects of global change", HRJRG-221). LaBeglo is funded by the Helmholtz Association and the Russian Foundation for Basic Research (RFBR). Further support was by research scholarships from the DAAD and the Russian Ministry of Science under the joint program "Mikhail Lomonosov," by the Russian Foundation for Basic Research (projects 14-04-00501, 15-04-06685), Russian Science Foundation (project 14-14-00400) and by the foundation of "Goszadanie" from the Ministry of Education and Science of Russia.

References

Bedulina, D.S., Evgen'ev, M.B., Timofeyev, M.A., Protopopova, M.V., Garbuz, D.G., Pavlichenko, V.V., Luckenbach, T., Shatilina, Z.M., Axenov-Gribanov, D.V., Gurkov, A.N., Sokolova, I.M., Zatsepina, O.G., 2013. Expression patterns and organization of the hsp70 genes correlate with thermotolerance in two congener endemic amphipod species (Eulimnogammarus cyaneus and E. verrucosus) from Lake Baikal. Mol Ecol 22, 1416-1430.

Bedulina, D.S., Takhteev, V.V., Pogrebnyak, S.G., Govorukhina, E.B., Madyarova, E.V., Lubyaga, Y.A., Vereshchagina, K.P., Timofeyev, M.A., Luckenbach, T., 2014. On Eulimnogammarus messerschmidtii, sp n. (Amphipoda: Gammaridea) from Lake Baikal, Siberia, with redescription of E. cyanoides (Sowinsky) and remarks on taxonomy of the genus Eulimnogammarus. Zootaxa 3838, 518-544.

Berezina, N.A., 2007. Invasions of alien amphipods (Amphipoda: Gammaridea) in aquatic ecosystems of North-Western Russia: pathways and consequences. Hydrobiologia 590, 15-29.

Christensen, J.H., Hewitson, B., Busuioc, A., Chen, A., Gao, X., Held, I., Jones, R., Kolli, R.K., Kwon, W.-T., Laprise, R., Magaña Rueda, V., Mearns, L., Menéndez, C.G., Räisänen, J., Rinke, A., Sarr, A., Whetton, P., 2007. Regional Climate Projections., in: Solomon, S., Qin, D., Manning, M., Chen, Z., Marquis, M., Averyt, K.B., Tignor, M., Miller, H.L. (Eds.), Climate Change 2007: The Physical Science Basis. Contribution of Working Group I to the Fourth Assessment Report of the Intergovernmental Panel on Climate Change. Cambridge University Press, Cambridge, United Kingdom and New York, NY, USA., pp. 847-940.

Clarke, A., Murphy, E.J., Meredith, M.P., King, J.C., Peck, L.S., Barnes, D.K., Smith, R.C., 2007. Climate change and the marine ecosystem of the western Antarctic Peninsula. Philos Trans R Soc Lond B Biol Sci 362, 149-166.

Falkner, K.K., Measures, C.I., Herbelin, S.E., Edmond, J.M., Weiss, R.F., 1991. The major and minor element geochemistry of Lake Baikal. Limnol Oceanogr 36, 413-423.

Gladyshev, M.I., Moskvicheva, A.V., 2002. Baikal invaders have become dominant in the upper Yenisei benthofauna. Doklady Biological Sciences 383, 138-140.

Grabowski, M., Bacela, K., Konopacka, A., 2007. How to be an invasive gammarid (Amphipoda: Gammaroidea) - comparison of life history traits. Hydrobiologia 590, 75-84.

Grigorovich, I.A., Angradi, T.R., Emery, E.B., Wooten, M.S., 2008. Invasion of the Upper Mississippi River system by saltwater amphipods. Fundam. Appl. Limnol., Arch. Hydrobiol. 173, 67-77.

Hampton, S.E., Izmest'eva, L.R., Moore, M.V., Katz, S.L., Dennis, B., Silow, E.A., 2008. Sixty years of environmental change in the world's largest freshwater lake - Lake Baikal, Siberia. Glob Change Biol 14, 1947-1958.

Hartung, T., Rovida, C., 2009. Chemical regulators have overreached. Nature 460, 1080-1081.

IPCC, 2007. Intergovernmental Panel on Climate Change (IPCC). Fourth Assessment Report. http://www.ipcc.ch/ipccreports/ar4-wg1.htm.

Ishibashi, H., Iwata, H., Kim, E.Y., Tao, L., Kannan, K., Amano, M., Miyazaki, N., Tanabe, S., Batoev, V.B., Petrov, E.A., 2008. Contamination and effects of perfluorochemicals in Baikal seal (Pusa sibirica). 1. Residue level, tissue distribution, and temporal trend. Environ Sci Technol 42, 2295-2301.

Karaman, G.S., Pinkster, S., 1977. Freshwater Gammarus species from Europe, North Africa and adjacent regions of Asia (Crustacea-Amphipoda). Part I. Gammarus pulex-group and related species. Bijdragen Tot De Dierkunde 47, 1-97.

Khozov, M., 1963. Lake Baikal and its life, The Hague, Netherlands.

Kozhova, O.M., Izmest'eva, L.R., 1998. Lake Baikal - Evolution and Biodiversity. Backhuys Publishers, Leiden.

Kravtsova, L.S., Izhboldina, L.A., Khanaev, I.V., Pomazkina, G.V., Rodionova, E.V., Domysheva, V.M., Sakirko, M.V., Tomberg, I.V., Kostornova, T.Y., Kravchenko, O.S., Kupchinsky, A.B., 2014. Nearshore benthic blooms of filamentous green algae in Lake Baikal. J Great Lakes Res 40, 441-448.

Lannig, G., Cherkasov, A.S., Portner, H.O., Bock, C., Sokolova, I.M., 2008. Cadmium-dependent oxygen limitation affects temperature tolerance in eastern oysters (Crassostrea virginica Gmelin). Am J Physiol Regul Integr Comp Physiol 294, R1338-1346.

Magnuson, J.J., Robertson, D.M., Benson, B.J., Wynne, R.H., Livingstone, D.M., Arai, T., Assel, R.A., Barry, R.G., Card, V.V., Kuusisto, E., Granin, N.G., Prowse, T.D., Stewart, K.M., Vuglinski, V.S., 2000. Historical trends in lake and river ice cover in the northern hemisphere. Science 289, 1743-1746.

Mamontov, A.A., Mamontova, E.A., Tarasova, E.N., McLachlan, M.S., 2000. Tracing the sources of PCDD/Fs and PCBs to Lake Baikal. Environ. Sci. Technol. 34, 741-747.

Mazepova, G., 1990. Rakushkovye rachki (ostracoda) Baykala (Ostracoda of Lake Baikal). Nauka, Novosibirsk.

Moore, M.V., Hampton, S.E., Izmest'eva, L.R., Silow, E.A., Peshkova, E.V., Pavlov, B.K., 2009. Climate Change and the World's "Sacred Sea"-Lake Baikal, Siberia. Bioscience 59, 405-417.

Pavlichenko, V., Protopopova, M., Timofeyev, M., Luckenbach, T., 2015. Identification of a putatively multixenobiotic resistance related Abcb1 transporter in amphipod species endemic to the highly pristine Lake Baikal. Environmental Science and Pollution Research, in press.

Protopopova, M.V., Pavlichenko, V.V., Menzel, R., Putschew, A., Luckenbach, T., Steinberg, C.E.W., 2014. Contrasting cellular stress responses of Baikalian and Palearctic amphipods upon exposure to humic substances:

environmental implications. Environmental Science and Pollution Research 21, 14124-14137.

Rivarola-Duarte, L., Otto, C., Jühling, F., Schreiber, S., Bedulina, D., Jakob, L., Gurkov, A., Axenov-Gribanov, D., Sahyoun, A.H., Lucassen, M., Hackermüller, J., Hoffmann, S., Sartoris, F., Pörtner, H.O., Timofeyev, M., Luckenbach, T., Stadler, P.F., 2014. A first glimpse at the genome of the Baikalian amphipod Eulimnogammarus verrucosus. J Exp Zool Part B 322, 177-189.

Rusinek, O.T., Takhteev, V.V., Gladkochub, D.P., Khodzher, T.V., Budnev, N.M., 2012a. Baicalogy. Book 1. Nauka, Novosibirsk.

Rusinek, O.T., Takhteev, V.V., Gladkochub, D.P., Khodzher, T.V., Budnev, N.M., 2012b. Baicalogy. Book 2. Nauka, Novosibirsk.

Shimaraev, M., Verbolov, V., Granin, N., Sherstayankin, P., 1994. Physical limnology of Lake Baikal: a review. BICER, Baikal International Center for Ecological Research.

Stachowicz, J.J., Terwin, J.R., Whitlatch, R.B., Osman, R.W., 2002. Linking climate change and biological invasions: Ocean warming facilitates nonindigenous species invasions. Proc Natl Acad Sci U S A 99, 15497-15500.

Stift, M., Michel, E., Sitnikova, T.Y., Mamonova, E.Y., Sherbakov, D.Y., 2004. Palaearctic gastropod gains a foothold in the dominion of endemics: range expansion and morphological change of Lymnaea (Radix) auricularia in Lake Baikal. Hydrobiologia 513, 101-108.

Takhteev, V.V., 2000. Essays on Lake Baikal´s amphipods: systematics, comparative ecology, evolution Irkutsk State University Press, Irkutsk.

Timofeyev, M.A., 2010. Ecophysiological aspects of adaptation to abiotic environmental factors in endemic Baikal and Palearctic amphipods. Irkutsk State University, Irkutsk, Russia.

Timoshkin, O., 1999. Biology of Lake Baikal: "White spots" and progress in research. Berliner Geowissenschaftliche Abhandlungen E 30, 333-348.

Timoshkin, O.A., Bondarenko, N.A., Volkova, E.A., Tomberg, I.V., V.S., V., Maljnik, V.V., 2014. Mass development of filamentous algae of genera Spirogyra and Stigeoclonium (Chlorophyta) in the coastal zone of southern Baikal. Hydrobiological Journal 50, 15-26 (In Russian).

Timoshkin, O.A., 2001. Lake Baikal: fauna diversity, problems of its "immiscibility" and origin, ecology and "exotic" communities., in: Timoshkin, O.A. (Ed.), Index of animal species inhabiting Lake Baikal and its catchment area. Nauka Publishers, Novosibirsk, pp. 16-73.

Van den Brink, F.W.B., Van der Velde, G., De Vaate, A.B., 1991. Amphipod invasion on the Rhine. Nature 352, 576-576.

Weinberg, I.V., Kamaltynov, R.M., 1998. Macrozoobenthic communities of rocky beach of Lake Baikal. 1. Fauna. Zoologicheskii zhurnal 77, 259-265 (In Russian).

Weiss, R.F., Carmack Carmack, E.C., Koropalov, V.M., 1991. Deep-water renewal and biological production in Lake Baikal. Nature 349, 665-669.

Yoshioka, T., Ueda, S., Khodzher, T., Bashenkhaeva, N., Korovyakova, I., Sorokovikova, L., Gorbunova, L., 2002. Distribution of dissolved organic carbon in Lake Baikal and its watershed. Limnology 3, 0159-0168.

The influence of BPPC on Baikal plankton – comparative study of phytoplankton in the point of influence of BPPC purified waste waters and in the reference clean point in 2005-2006 years

Svetlana V. Shimaraeva, Lyubov R. Izmestyeva, Lyudmila S. Krashchuk, Helene V. Pislegina, Eugene A. Silow

Institute of Biology of Irkutsk State University, Irkutsk-3, POBox 24, 664003, Russia
shimaraeva@gmail.com

Abstract

This work demonstrates principal similarity of the processes occurring in phytoplankton communities near western shore (at reference clean "point #1") and near eastern shore of Southern Baikal (at influenced by purified waste waters point "testing area P7") in June-October of 2005-2006. Though in both years phytoplankton community at clean "point #1" was characterized by higher biodiversity, principal composition of phytoplankton was similar and processes taking place in both points resemble each other. The year 2005 can be counted as "phytoplankton rich development year", while the following year 2006 can be named as "phytoplankton medium development year". Our estimates can conclude stable equilibrium state of autotrophic chain of Southern Baikal ecosystem.

Introduction

Institute of Biology fulfills the monitoring observations after the state of biological communities in the point of possible influence of purified waste waters of Baikalsk Pulp and Paper Combine (BPPC) starting from the time, when its building has been planned (1960) (Kozhova, Izmestyeva, 1998), comparing these results with the reference clean "point #1", situated near the opposite shore of the lake, where monitoring started in 1946 (Kozhov, 1963).

The years described here were among the last years, when Combine functioned for the full power, as starting with 2008 it stops from time to time and has been turned out completely in 2013.

The aim of our recent work – comparison of phytoplankton development during Summer-Autumn periods in two sites: one – influenced by BPPC activity and another – situated in far from BPPC location, both belonging to Southern Baikal and situated in the similar geographic, natural, climatic etc. conditions.

We have concentrated our attention on the assessment of the state of phytoplankton, as it is the main component determining the quality of aquatic community, providing the most energy for the ecosystem as well as the main part of biological matter turnover.

Sites of and methods of observations

Reference clean "point #1": It is situated near the western coast of Southern Baikal, in 2.7 km from the shore in front of biological station of Institute of Biology with the depth of ≈ 800 m, geographical coordinates are N – 51° 54'.195, E – 105° 04'.235. This site is shown to reflect the situation in Southern Baikal adequately (Kozhov, 1963; Kozhova, Izmestyeva, 1998).

Influenced by purified waste waters point "testing area P7": is situated near the eastern coast of Southern Baikal, abreast of injector of discharge of purified waste waters of BPPC, in 7 km from the shore with the depth of ≈ 900 m, geographical coordinates are N – 51° 33'.19, E – 104° 19'.5.

Water to analyze chlorophyll a content and determination of number and species composition of phytoplankton was sampled with 7 l Van Dorn bottle (bathometer) from the depths of 0, 10, 25, 50, 100 m. Water temperature was measured with mercury thermometer inserted in the bathometer, water transparency – with white Secci disk, 35 cm in diameter. Chlorophyll a concentration was determined with standard spectrophotometric method (Report..., 1964) after filtration of water through nucleopore filters with pores of diameter of 0.7 µm.Phytoplankton samples were fixed with the Utermöhl solution, settled for 2 weeks and analyzed with light microscope (256 forms of algae were distinguished and counted).

Results for 2005

State of phytoplankton at reference clean "point #1"

Surface temperature in beginning of Summer was 3.7 °C (June 2005), then water started to warm in the beginning of July and in the beginning of August reached maximum (15.1 °C in the surface layer). The warming period covered total trophogenic layer till the end of September.

Weighted average temperature in the layer 0-50 m was 10.8 °C. Water transparency in 2005 varied from 4 m (end of July) till 21 m (October). Chlorophyll a content in 0-50 m layer in July-October varied widely. The total series is characterized by the following data:

Average	1.04	Excess	-0.42
Standard error	0.14	Asymmetry	-0.69
Median	1,10	Interval	1.58
Standard deviation	0.49	Minimum	0.09
Sampling variance	0.24	Maximum	1.66

Phytoplankton composition in July-October, 2005 г. is characterized by the presence of 60 algae taxa, belonging to 6 orders.

The most varied were **Bacillariophyta** – we've met 24 species, not taking into account benthic forms. Nevertheless, their number was law – maximal numbers varied within from 0.02 10^3 cells l^{-1} for *Aulacoseira skvortzowii* to 30.12 10^3 cells l^{-1} for *Stephanodiscus* sp. Number of species of **green** algae was 19. Their number varies in wide limits – from 0.16 10^3 cells l^{-1} for *Oocystis parva* to 42864 10^3 cells l^{-1} for *Pseudoictyosphaerium minusculum*. **Cyanophyta** were presented by 7 species. Their maximal number lied within from 2.04 10^3 cells l^{-1} for *Anabaena* sp. to 189410.00 10^3 cells l^{-1} for *Synechocystis limnetica*. **Chrysophyta** were also presented by 7 species with oscillations of number from 1.15 10^3 cells l^{-1} for *Dinobryon elegantissimum* to 624.03 10^3 cells l^{-1} for *Chrysochromulina* sp. Two species of **Cryptophyta** were registered: *Chroomonas acuta* and *Chroomonas* sp. with numbers 807.22 and 49.47 10^3 cells l^{-1}, corresspondingly. **Dinophyta** were presented also by two species:

Gymnodinium coeruleum and *Glenodinium* sp. sp. with maximal number of 3.71 and 2.82 10^3 cells l^{-1}.

In more number – maximums of 216.85 and 749.37 10^3 cells l^{-1}, we met spherical non-identified algae and non-identified flagellata. Small-cell pico- and nanoplanktonic forms dominated, reaching up to 200 10^6 cells l^{-1} in some days. There were non-identified single cocci (1-2 µm in diameter), cocci (3–5 µm in diameter), blue-green: Baikal endemic *Synechocystis limnetica* and *Cyanodictyon planctonicum*. In mass developed flagellates (in some days their number could reach 200 10^3 cells l^{-1}). So, the list of dominant species (with maximal number of 100 10^3 cells l^{-1} and more) had the following composition (table 1):

Chroomonas acuta	807.22
Chrysidalis sp.	542.38
Dinobryon sociale var. sociale	384.46
Ankistrodesmus pseudomirabilis	278.67
Chrysochromulina sp.	219.15

Table 1: Dominant species of phytoplankton number (103 cells l-1) at reference clean "point #1" in 2005

Total number of phytoplankton (without taking into account nano–, picoplankton and flagellates) started to rise rapidly from the middle of July, reached its maximum (weighted average in 0-50 m more than 825 10^3 cells l^{-1}) in August 20th, remained high for total September and beginning of October, and declined in the end of October.

Total number of phytoplankton including nano–, picoplankton and flagellates was maximal in the first decade of August (weighted average in 0–50 m was more than 90 10^6 cells l^{-1}). So, 2005 Summer we can call phytoplankton-reach year.

State of phytoplankton at influenced by purified waste waters point "testing area P7"

Water masses near eastern shore was also well warmed in 2005. Weighted average temperate for 0-50 m were some lower than near western shore in August, the same in September and some even higher in October. In October temperature at 50 m depth was still not lower 6 °C. Water transparency was about 10 m and was not greatly different from transparency at western shore. Chlorophyll *a* content at influenced by purified waste waters point "testing area P7" practically not differed from the one at "point #1" in 2005, July-October. Maximal concentrations are registered in August at maximal water temperature.

The total data are not significantly different from the "point #1 and they can be counted as the same general population:

Average	1.12	Excess	0.72
Standard error	0.04	Asymmetry	0.74
Median	1,10	Interval	3.17
Standard deviation	0.63	Minimum	0.05
Sampling variance	0.40	Maximum	3.22

We've discovered 45 algae taxa in "testing area P7", that is sufficiently less than at "point #1".

Among them **Bacillariophyta** – 7 species with maximal number from 0.02 to 70.51 10^3 cells l^{-1}. Endemics of Baikal – two species only *Aulacoseira skvortzowii* and *Stephanodiscus skabitschevskii* with insignificant number of less than 1 10^3 cells l^{-1}. **Green algae** - 11 species, with maximal number from 0.15 to 17716 10^3 cells l^{-1}. **Chrysophyta** - 7 species with maximal number from 0.09 до 625,62 10^3 cells l^{-1}. **Dinophyta** – 2 species with maximal number less than 2 10^3 cells l^{-1}. 2 species of **blue-green alga**, with endemic of Baikal *Synechocystis limnetica* (60.12 10^3 cells l^{-1}) and falciform non-identified alga (16.37 10^3 cells l^{-1}). **Flagellates** were multiple and various – 8 species with maximal number from 6.54 to 309.06 10^3 cells l^{-1}. The most dominating forms were small-cell pico– and nanoplanktonic alga, flagellates. Their total number some days was close to 100 10^6 cells l^{-1}. There were non-identified single cocci (1-2 μm in diameter), cocci (3–5 μm in diameter), blue-green: Baikal endemic *Synechocystis limnetica* and *Cyanodictyon planctonicum,* green *Pseudoictyosphaerium minusculum.* In August dominant species were represented (without small-cell) by green and Chrysophyta (table 2):

Chroomonas acuta	241.11
Dinobryon bavaricum var. *bavaricum*	199.98
Ankistrodesmus pseudomirabilis	65.71
Chrysidalis sp.	62.51
Dinobryon sociale var. *sociale*	32.30

Table 2: Dominant species of phytoplankton number (103 cells l-1) at "testing area P7" in August 2005

In September dinobryons were lost from the list of dominant species, while green *Ankistrodesmus angustus* and diatom *Cyclotella minuta* were included (table 3):

Chrysidalis sp.	324.79
Chroomonas acuta	211.67
Ankistrodesmus pseudomirabilis	48.41
Ankistrodesmus angustus	21.57
Cyclotella minuta	17.82

Table 3: Dominant species of phytoplankton number (103 cells l-1) at "testing area P7" in September 2005

In October the list of dominant forms was as follows (table 4):

Chrysidalis sp.	625.62
Chroomonas acuta	286.23
Cyclotella minuta	70.51
Stephanodiscus hantzschii var. *pusillus*	69.64
Ankistrodesmus pseudomirabilis	33.32

Table 4: Dominant species of phytoplankton number (10^3 cells l^{-1}) at "testing area P7" in October 2005

As we can see, one more diatom *Stephanodiscus hantzschii* var. *pusillus* joined to dominants, *Ankistrodesmus angustus* was lost, and maximal number of algae has increased when compared to September.

Generally speaking the list of dominant species (exceeding 100×10^3 cells l^{-1}) at "testing area P7" during Summer-Autumn 2005 contained the following species (see table 5):

Chroomonas acuta	286.23
Dinobryon bavaricum var. *bavaricum*	199.98
Chrysidalis sp.	625.62

Table 5: Dominant species of phytoplankton number (10^3 cells l^{-1}) at "testing area P7" in 2005

Total number of phytoplankton, including small-cell forms, was maximal in August – weighted average in 0-50 m reached 50339 10^3 cells l^{-1}.

Results for 2006

State of phytoplankton at reference clean "point #1"

Content of chlorophyll a in 0-50 m layer in July-October 2006 varied from 0.01 to 3.83 мg м$^{-3}$. Weighted average series was characterized by the following parameters:

Average	1.05	Excess	−1.56
Standard error	0.18	Asymmetry	0.04
Median	1.17	Interval	2.10
Standard deviation	0.72	Minimum	0.07
Sampling variance	0.51	Maximum	2.17

Content of chlorophyll *a* was closely related with water transparency like in 2005. About 70 taxa of alga, belonging to 6 orders, were met at "point #1" in June-October 2006.

The most multiple and diverse group was **Bacillariophyta** – we have counted 28 species (not taking into account benthic non-identified forms). The maximal numbers lied within limits from 0.02 10^3 cells l^{-1} for *Aulacoseira skvortzowii* to 205.96 10^3 cells l^{-1} for *Stephanodiscus hantzschii* var. *pusillus.*
There were 24 species of **green algae**. Their maximal number varied from 0.07 10^3 cells l^{-1} for *Schroederia setigera* to 155.95 10^3 cells l^{-1} for *Ankistrodesmus pseudomirabilis.* There were 9 species belonging to **Cyanophyta**. Their maximal number varied greatly from 0.07 10^3 cells l^{-1} for *Aphanizomenon flos-aquae* f. *flos-aquae* to 19887.13 10^3 cells l^{-1} for *Synechocystis limnetica.*
Chrysophyta were presented by 10 species with maximal number from 1.09 for *Dinobryon divergens* var. *divergens* to 352.82 10^3 cells l^{-1} for *Chrysidalis* sp. Traditionally there were only two **Cryptophyta** species – *Chroomonas acuta* and *Chroomonas* sp. with maximal numbers of 471.24 10^3 cells l^{-1} and 21.91 10^3 cells l^{-1}. **Dinophyta** were presented by three species *Gymnodinium coeruleum Glenodinium* sp. sp., and *Glenodinium* «mushrooomlike» with maximal numbers of 5.35 10^3 cells l^{-1}, 5.16 10^3 cells l^{-1} and 2.75 10^3 cells l^{-1} consequently. Non-identified spheric cells and flagellates were met in higher maximal numbers 2052.63 10^3 cells l^{-1} and 227.99 10^3 cells l^{-1} consequently.

For each month we have created list of the first ten most multiple species (without nanoplankton forms). In June they were: *Chroomonas acuta, Ankistrodesmus pseudomirabilis, Chroomonas* sp., *Chrysidalis* sp., *Chromulina* sp., *Stephanodiscus dubius, Cyclotella minuta, Ankistrodesmus angustus, Glenodinium* sp. sp. и *Lambertia (Ankyra).* Variety of maximal number was rather significant: from 1.78 10^3 cells l^{-1} for *Lambertia (Ankyra)* to 289.02 10^3 cells l⁻¹ for *Chroomonas acuta*. In July the list changed by 60%. *Chroomonas acuta, Ankistrodesmus pseudomirabilis, Chroomonas* sp., *Chrysidalis* sp. remained in the list, with addition of *Nitzschia acicularis, Stephanodiscus dubius, Stephanodiscus* sp., *Synedra acus, Koliella longiseta* f. *longiseta, Stephanodiscus hantzschii* var. *pusillus*. In August to *Chroomonas acuta, Chrysidalis* sp., *Ankistrodesmus pseudomirabilis, Stephanodiscus hantzschii* var. *pusillus, Stephanodiscus* sp., 2 Chrysophyta species: *Dinobryon sociale* var. *sociale (D. sociale tipica)* and *D. sociale* var. *stipitatum*, 2 species of blue-green *Anabaena Lemmermannii, A. flos-aquae* and green *Ankistrodesmus minutissimus* joined the list. In September diatom *Cyclotella minuta,* and *Ankistrodesmus angustus, Ankistrodesmus* sp. и *Chroomonas* sp. joined this list.

The total level of phytoplankton development remained high, as the number of three species was higher than 100 10³ cells l⁻¹.We observed in September the highest number of *Chroomonas acuta* and *Chrysidalis* sp., 471.24 10³ cells l⁻¹ and 352.82 10³ cells l⁻¹. In October list remained similar to September one. 2 species of green *Ankistrodesmus*, were replaced by *Didymocystis* sp. and *Dinobryon cylindricum* var. *cylindricum*.

So, list of mass species, met in June-October in the number of more 100 10³ cells l⁻¹, included the following forms (table 6):

Chroomonas acuta	471.24
Chrysidalis sp.	352.82
Stephanodiscus hantzschii var. *pusillus*	205.96
Stephanodiscus sp.	185.52
Ankistrodesmus pseudomirabilis	155.95
Dinobryon sociale var. *sociale*	105.27

Table 6: Dominant species of phytoplankton number (10³ cells l⁻¹) at "point #1" in 2006

We can see the change of mass species composition. There is not *Chrysochromulina* sp. in this species, and there are 2 species *Stephanodiscus hantzschii* var. *pusillus* и *Stephanodiscus* sp., not developed in mass in 2006.

The basis of phytoplankton community (from 02.06. to 11.07.2006) was *Chroomonas acuta*. At the depth of 10 m, where the number of algae is maximal, the number of this species was more than 50 % of total phytoplabkton number. It was accompanied by green *Ankistrodesmus pseudomirabilis* and other species of genus *Chroomonas*. Generally speaking we can count phytoplankton community of this time monodominant. In the end of July-August the share of *Chroomonas acuta* is decreased, but remains relatively high and this species did not lost his position among dominants. *Ankistrodesmus pseudomirabilis* joined the list of co-dominants, together with *Nitzschia acicularis,* in August - *Stephanodiscus,* in September - *Chrysidalis* sp. And in October *Cyclotella minuta* joined co-dominant team.

So, in August-September phytoplankton community becomes more diverse, the share of super-dominant *Chroomonas acuta* decreases, the role of other species increases, community transfers to polydominant.

According to number, phytoplankton was dominated by small-cell pico- and nanoplankton algae. Their total number reached sometimes 267 10^6 cells l^{-1}. There were non-identified single cocci (1-2 μm in diameter), cocci (3–5 μm in diameter), blue-green: Baikal endemic *Synechocystis limnetica* and *Cyanodictyon planctonicum*. There was mass development of non-identified small flagellates 3–8 μm and other flagellates – their number reached 1966.77 10^3 cells l^{-1} and 2052.63 10^3 cells l^{-1} consequently. Seasonal dynamics of small-cell algae repeated one of dominant phytoplankton. The maximal numbers were registered in August.

Weighted average number of phytoplankton in June-October 2006 in 0-50 m varied from 106 10^3 cells l^{-1} to 938 10^3 cells l^{-1}. Characteristics of it were the following:

Average	307.34	Excess	5.38
Standard error	51.49	Asymmetry	2.10
Median	247.82	Interval	831.56
Standard deviation	205.98	Minimum	106.24
Sampling variance	42428	Maximum	937.80
Number of observations	16		

Baikal endemic species of algae whether were not met in June-October, 2006, or were met in very small number (Table 7)

Aulacoseira baicalensis	0.20
Aulacoseira skvortzowii	0.15
Cyclotella baicalensis	0.16
Cyclotella minuta	25.63
Synechocystis limnetica	19887,13

Table 7: Maximal registered number of Baikal endemic species of algae (10^3 cells l^{-1}) at at reference clean "point #1" in June-October, 2006

It is clear seen the mass development of just one blue-green small-cell species *Synechocystis limnetica*. The lack or small quantity of other species can be explained by the fact they prefer under-ice season.

State of phytoplankton at influenced by purified waste waters point "testing area P7"

In 2006 at "testing area P7" the temperature of surface layer was significantly higher than at "point #1". In July and September it was higher 12,6 и 12,5 °C, comparing the temperature at "point #1" - 5,8 и 9,9 °C, consequently. Nevertheless, weighted average temperatures for "testing area P7" and "point #1" for 0-50 m were not so significantly different – 5,2 °C (at "P7") and 4,9 °C (at "point #1").

Transparency at stations "P7" and"#1"were different in July - 8.5 m and 15.5 m, but in September they were 8.5 and 7.0 м, consequently. Chlorophyll a content in July and September at "P7" was some higher than "#1". Weighted average chlorophyll a content for 0–100 m at influenced by purified waste waters point "testing area P7" were 0.64 mg m^{-3} in July and 0.91 mg m^{-3} in September. In July-September 2006 at purified waste waters point "testing area P7" we registered 19 taxa of algae.

There were 4 species of **diatoms** with maximal numbers from 0.41 10^3 cells l^{-1} to 10.66 10^3 cells l^{-1}, including endemics of Baikal: *Cyclotella minuta* - 4,20 10^3 cells l^{-1}, *Aulacoseira baicalensis* and *Cyclotella baicalensis* with number less than 1 10^3 cells l^{-1}. Among **green** algae (3 species), maximal numbers varied from 0.16 10^3 cells l^{-1} to 27.24 10^3 cells l^{-1}. **Chrysophyta** (also 3 species) with maximal numbers from 0.55 10^3 cells l^{-1} to 480.44 10^3 cells l^{-1}.

2 species of **Dinophyta** with maximal numbers from 1.59 to 9.23 10^3 cells l^{-1}.

Cryptophyta – 1 species with maximal number 900.94 10^3 cells l^{-1}. 1 species of **blue-green** – endemic of Baikal *Synechocystis limnetica* with maximal number 12628.00 10^3 cells l^{-1}. Maximal number of **flagellates** (2 species), varied from 1.48 to 9.66 10^3 cells l^{-1}.Non-identified **spheric algae** with maximal number from 1.63 to 48510.00 10^3 cells l^{-1}.

The most multiple group in phytoplankton as usually was presented by small-cell pico- and nanoplankton forms. Their summary number some days was about 60 10^6 cells l^{-1}. There were non-identified single cocci (1-2 μm in diameter), and blue-green Baikal endemic *Synechocystis limnetica*.In July dominant group consisted of (without small-cell) (10^3 cells l^{-1}):

Chroomonas acuta	900.94
Chrysidalis sp.	480.44

In September the list of dominant forms was as follows (10^3 cells l^{-1}):

Chroomonas acuta	283.75
Chrysidalis sp.	119.36
Ankistrodesmus pseudomirabilis	27.24
Sphaerocystis planctonica	23.87

For Summer-Autumn 2006 dominant algae (with number more than 100 10^3 cells l^{-1}) for purified waste waters point "testing area P7" were:

Chroomonas acuta	900,94
Chrysidalis sp.	480,44

Total number of phytoplankton, including small-cell forms was maximal in July – weighted average for 0-50 m layer was equal to 29999,28 10^3 cells l^{-1}.The number of large-cell phytoplankton in July at eastern shore was higher than at western one. At depth 10 m (maximal number of phytoplankton) the number was 5 times higher than at western shore. The composition of dominant species also differed – at western shore it was more diverse, though in both cases the main dominant was *Chroomonas acuta*. Superdominant of Southern Baikal was small-cell endemic **Cyanophyta** *Synechocystis limnetica*.

Phytoplankton community at eastern shore of Southern Baikal was analogous to that of 2005. This group consisted from Cryptophyta, Chrysophyta and Chlorophyta: *Chroomonas acuta, Chrysidalis* sp. и *Ankistrodesmus pseudomirabilis*. Baikal endemic from diatom group presented in phytoplankton, but in low number, that is typical for this season.

Discussion

We can see that in July-October 2005 variability of chlorophyll *a* content in photic layer was within typical for Baikal limits and did not differ significantly at western and eastern shores. In June-October of 2006 the variability of chlorophyll *a* content of Southern Baikal remained within the typical for the Lake Baikal limits and has no significant differences for eastern and western shores.

The super-dominants for phytoplankton of Southern Baikal were small-cell algae belonging to blue-green and green orders, including endemic *Synechocystis limnetica*. According to degree of Summer plankton development the year 2005 may be called "rich in alga". Degree of development of Summer phytoplankton in 2006 can be assessed as "medium". This is equally veracious for pelagic waters of both western and eastern shores.

Judging by species richness algal diversity was higher at western shore, than at eastern one in 2005. Here we counted 60 algal taxa, while at eastern shore 45 only. The main composition of phytoplankton community at western shore, as well as at eastern shore was determined by Chrysophyta, Cryptophyta and green alga: *Chroomonas acuta, Chrysidalis* sp., *Dinobryon sociale sociale* and *Ankistrodesmus pseudomirabilis.* Baikal endemics from diatoms were present in both regions, but were not very multiple, which is typical for this season (Izmestyeva et al., 2011). In October, with cooling of water mass endemic diatom *Cyclotella minuta* started to join to dominant community.

Main dominants of Southern Baikal phytoplankton were small-cell algae belonging to **Cyanophyta** and **Chlorophyta,** including endemic species *Synechocystis limnetica.* Quantitative characteristics of phytoplankton development in 2006 were lower than for 2005, including large-cell plankton, and total phytoplankton (together with small-cell phytoplankton).

By species diversity phytoplankton community at western shore in 2006 was richer (70 forms) than at eastern one (19 forms), more than in 2005. The composition of dominant complex was typical for Baikal for this season and was similar to 2005. The composition of phytoplankton community was determined by Cryptophyta and Chrysohyta – *Chroomonas acuta, Chrysidalis* sp. Endemics of Baikal belonging to Bacillariophyta presented in both studied regions, but were in low number, which is typical for this season (Izmestyeva et al., 2011). In September-October with cooling of water masses, endemic diatom *Cyclotella minuta* appeared among dominant forms.

These estimates can conclude stable equilibrium state of autotrophic chain of Southern Baikal ecosystem. The changes observed are of natural character.

Acknowledgements

This work was partly supported by Russian Ministry of of Education and Science (R212-IB-001).

Conclusion

We demonstrated principal similarity of the processes occurring in phytoplankton communities near western shore (at reference clean "point #1") and near eastern shore of Southern Baikal (at influenced by purified waste waters point "testing area P7"). Though in both years phytoplankton community at clean "point #1" was characterized by higher biodiversity, principal composition of phytoplankton was similar and processes taking place in both points resemble each other.

The year 2005 can be counted as "phytoplankton rich development year", while the following year 2006 can be named as "phytoplankton medium development year".

Our estimates can conclude stable equilibrium state of autotrophic chain of Southern Baikal ecosystem. The changes observed are of natural character, though with the same probability they can be explained by global, or local climate changes, or can be partly related with the human induced influence.

References

Izmestyeva LR, Silow EA, Litchman E (2011): Long-Term Dynamics of Lake Baikal Pelagic Phytoplankton under Climate Change. Inland Water Biology. 4(3): 301–307.

Kozhov MM (1963): Lake Baikal and its Life. The Hague: W.Junk. 344p.

Kozhova OM & Izmestyeva LR (Eds.) (1998): Lake Baikal. Evolution and Biodiversity. Leiden: Backhuys Publisher. 447 p.

Report of SCOR–UNESCO working group 17 (1964): Determination of photosynthetic pigments. Paris: UNESCO. 12 p.

V. Water management

Floods in the Selenga River basin: research experience

Garmayev Endon, Borisova Tatiana, Ayurzhanayev Alexander, Tsydypov Bair

The Baikal Institute of Nature Management, Ulan-Ude, Russia

Relevance

Flood is one of the most common hazardous disasters in the world that threatens human life and bears significant economic damage. Recent events in Russia, that have created the Federal Emergency (state) level in such rivers as the Amur, the Lena, the Kuban as well as in Krymsk, Altai and other regions, indicate a lack of hydrological study of a number of potentially dangerous rivers. In addition, the analysis of chronology of each element shows unwillingness and inability to fully appropriate structures for the prevention and prediction of catastrophic situations, protect the population from adverse impacts of water. For example, we note that because of the catastrophic flood on the Amur River, three regions of the Far East of Russia and northeastern provinces of China have suffered. The damages were enormous and were of hundreds of billions of rubles.

Thus, today, a detailed examination of major river basins and identifying patterns of floods development, inundation areas, possible damage, technical condition of existing defenses and substantiation for building the new ones seems to be necessary to solve this problem. However, such modern technological solutions as computerization of the received information with the possibility of modelling the process in its development will be the objective information base for management and operational competent decision-making to reduce social and economic losses, to prevent them in time and contribute to prospective development of the territory.

The Selenga River basin was chosen to be a testing ground for investigation as the most populous and developed part of the region. It is the largest river flowing into the lake, which goes through the territories of the two states: Russia (Buryatia) - Mongolia, with the catchment area of 447 thousand-km^2.

The basin area is a part of a central and buffer ecological zones of the Baikal natural territory with strict environmental limitations according to international standards. Its main tributaries are the Orkhon, the Eg (Mongolia), the Chikoi, the Khilok, the Uda, the Dzhida.

The Selenga refers to the rivers with a high probability of catastrophic freshets. Over the past 100 years, a series of major floods was seen on the Selenga: 02.07.1908; 11.08.1932; 06.11.1936; 08.05.1940; 08.05.1971; 29.07.1973 and a number of significant - 1931, 1938, 1942 and 1990. For example, the catastrophic floods in 1971 and 1973 paralyzed the republic completely: the airport was closed, the left bank part of the city of Ulan-Ude was cut off, connection with many areas was broken, bridges, roads, communication lines and power were damaged, a part of the population became homeless, an additional cleaning of the water was required, agricultural industry suffered huge losses. Those floods were classified as natural disasters on a national scale.

According to official statistics, the amount of damage inflicted to Buryatia reached: in 1971 - 1.4 billion Rubles, 1973 - 0.7 billion Rubles, 1993 - 40 billion Rubles (prices during the periods of floods). These figures reflect only the direct loss to the economy of the republic without significant environmental damage caused by water pollution from flooding and scour dumps, waste, cemeteries, animal burial grounds, etc. We should pay special attention to the fact that the Selenga River, carrying its waters into Lake Baikal, subjects to a high risk of environmental cleanliness and safety of the lake during the floods.

Thus, for a comprehensive study of floods and strategies to control them, solutions to the following problems are needed:

1. On the basis of mathematical modeling and analysis of long-term flow fluctuations of the transboundary Selenga River to determine the parameters of maximum (spring high water and rain freshets) river flow on locations of settlements on the territory of the Republic of Buryatia.

2. To develop evidence-based recommendations and interventions to prevent adverse impacts of the Selenga River waters; for flood protection of these settlements in the Republic of Buryatia.

Background information

At the initial stage in the dynamics for the period of 1936 - 2011, a collection of hydrographic parameters, data level mode, the maximum expenses, etc. was made, which was required for the subsequent operations. A unified database of diverse hydrological information, including the Mongolian part, was developed. Systematization of the entire volume of received information has allowed considering the formation and genesis of flooding in the space-time range.

In general, for the Selenga Basin rivers, raising of water levels and flows is typical during the periods of ice motion and the formation of ice jams, spring high waters from melting snow and glaciers, summer rain freshets.

Breakup on the rivers is often accompanied with ice jam phenomena, leading to a sharp short-term rise of water. Such local floods are confined to certain parts of the restrictions of riverbeds or bends. Spring high waters begin in April and usually take one wave. For the Selenga River high water is not typical. Rain freshets occur on the decline of high water and take place throughout the whole summer. High freshets in a year are usually seen in July - August. The maximum intensity of levels rise was observed on the Selenga Basin rivers, so during the passage of the highest freshet over 70 years on the Dzhida River (1971) it was 4.57 m/day (Khamenei village) and 2.79 m/day (Dzhida village). At the Chikoi River (1973) - 1.88 m/day (Povorot village) and 1.19 m/day (Cheremkhovo village); on the Uda River (1991) - 1.1 m/day. The decline is getting slower on the average 0.3-0.5 m/day. Besides, the rapid rise of levels of a number of mountain tributaries (the Khamenei, the Kurba, the Ona etc.) is also associated with their location in the zone of permafrost, which significantly weakens the infiltration capacity of soils.

The analysis of repeatability in their genesis indicates the predominance of freshet floods on the rivers of the Selenga River basin (61- 90%) and slightly – high water freshet floods (to 10%) The exception is the Uda River Basin, where small high water freshet floods are more frequent due to mountain rivers: the Kurba, the Ona. Floods caused by ice jams are possible in local areas of natural and man-made rivers clamps, restrictions and channels: the Khilok, the Selenga and rarely hanging ice dams - in the areas of intensive frazilization (the Selenga river – Murzino villiage, the Dzhida river – Dzhida station etc.) (Figure 1)

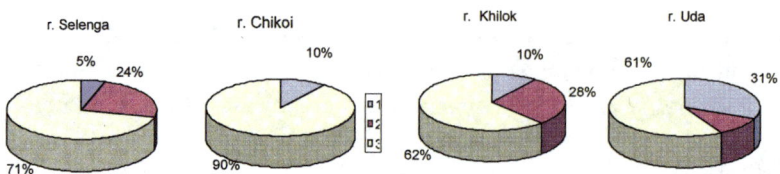

Fig. 1. A share of flooding kinds on the rivers of the Selenga river basin, %. 1 - high-water-freshet; 2 - jams; 3 - freshet.

Freshet floods are the most catastrophic. They are widespread, high-speed forming and can cover large areas.

Under ordinary floods, floodplain inundation depth does not exceed 0.5 - 1 m, and at higher floods reaches 1.8 - 3 m. The exceptional level rises above the critical (water outlet to the flood plain) were registered on separate hydrological stations: Dzhida - Khamenei (437 cm), Selenga - Novoselenginsk (419 cm), Chikoi - Povorot (267 cm), Uda - Ulan-Ude (266 cm). The height of the layer of water on the floodplain depends both on the strength of floods and hydrological and morphological characteristics of the river, so during floods on the Selenga river in the village of Ust-Kyakhta, it is 1 - up to 2 m, in Novoselenginsk village, under the conditions of a narrowing of the valley and a significant water supply from Dzhida and Chikoy rivers, it sharply increases and can reach more than 4 m, in Ulan-Ude it decreases to 2.2 m, and in the vast delta - up to 1 m.

The duration of high levels stand is different. Prolonged water whelms on floodplains (25-40 days) are observed in the valley of the Selenga River and in the lower reach of the Chikoi River. Shorter floods (up to 25 days) are found in the basins of the Barguzin, The Upper Angara, the Uda, the Dzhida and other watercourses. On small mountain rivers, usually it is not more than 3 - 7 days.

Based on generalization of the library, reference materials of the detailed reconnaissance survey, roster of settlements on the territory of the Selenga river basin, located in a flood zone, was composed. For example, a flood zone of the Selenga River covers more than 54 settlements, including the city of Ulan-Ude, 19 of them are located in the most hazardous areas. Inventory of existing hydraulic structures showed that the majority of them are not in a satisfactory condition and require urgent repair or reconstruction.

Results and consideration

The main result is the modeling of flood zones within the settlements at a given level of flood occurrence. It was performed using HEC-RAS, HEC-GeoRAS and ArcGIS software packages. The necessary primary data: 1. the creation of a digital elevation model of topographic maps of 1: 25,000 (for Ulan-Ude 1: 2000) for classical technology "Scanning - vectorization - digital map"; 2. The morphometric characteristics of river channels in DEM set according to pilot charts and depth sounder shooting; 3. Setting the geometric and morphometric characteristics of the channel.

The initial data for the construction of the relief was provided by the results of hydromorphological survey of the areas with broken and leveling cross sections on a channel and floodplain. The construction of profiles in relief was made with AutoCad software environment 2004 in two stages (Fig. 2).

Calculated water levels in cross sections of the settlements, obtained by the curves of water levels and flows occurrence in the gauges on the affected streams using graphs $Q = f(H)$, were used as hydrological information. For the settlements considered in this project, which do not have gauging stations, the method of interpolation with the longitudinal profile of the river and fall of stream based on large-scale maps was applied. As a result, the flood zones were identified - when the water level of summer freshets is of 1%, 5% occurrence, as well as the maximum level at ice dam phenomena.

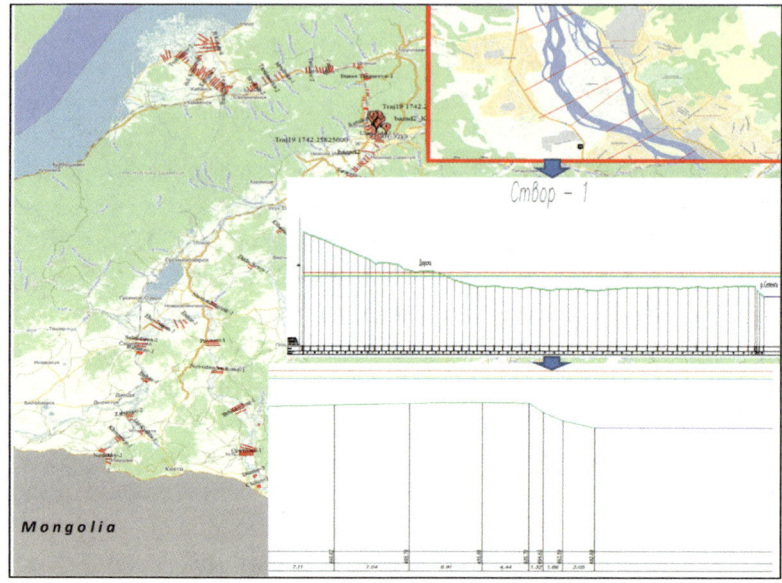

Figure 2. Hydromorphological survey. Location of the transverse profiles in the valleys of the Selenga and Chikoi rivers

Next, let us imagine the course of solving this problem: defining boundary conditions of model - design calculated water flow, Manning roughness coefficients for the channel and floodplain parts depending on the type of the underlying surface; solution in the HEC-RAS program of the simplified one-dimensional shallow water equations (Saint-Venant equations) using implicit finite-difference scheme. The eventual result is the level of the water surface in the calculated cross-sections in the period of a flood.

The resulting calculation data exported to the ArcGIS environment, where there is an automatic drawing of flooded areas on a digital elevation model. For example, in Figure 3 there is a flood zone of Ulan-Ude.

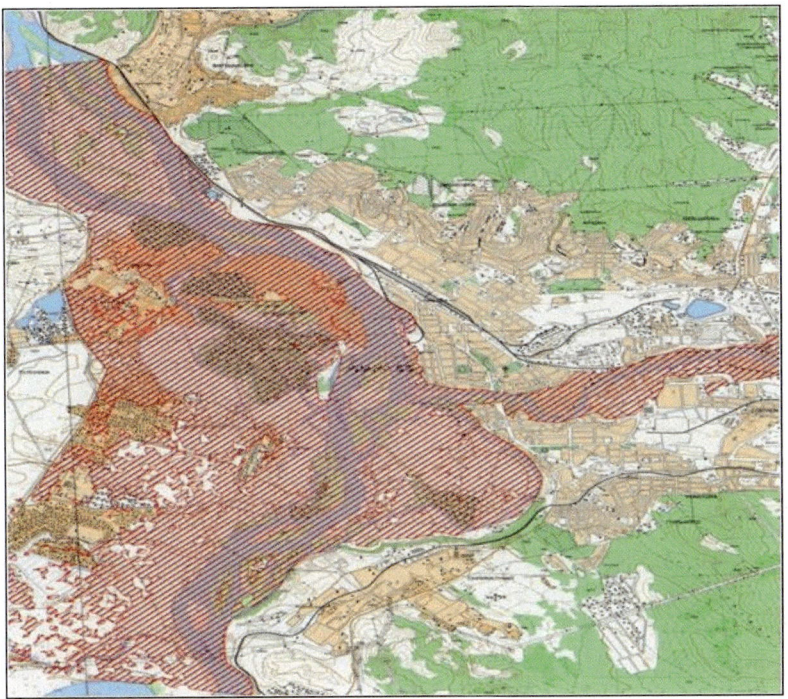

Figure 3. Flood zone of 1% occurring floods in Ulan Ude

The cartographic survey mapping of flood zones allowed estimating the extent of the negative impact of the water within the settlements, to determine the parameters and a list of objects and calculate probabilistic damage.

The analysis shows that the main problems are associated with the flooding of residential and commercial facilities, roads, communications, and agricultural land. Thus, when the freshet flood is of 1% occurrence, the area of possible flooding in residential areas is of 4878.54 ha and of maximum ice dam floods - 233.24 ha. The processes of coast collapse are most intensively developed within 24 settlements; during 25 years 142.3 hectares of land of the settlements may be lost. The hazardous areas are home to more than 39,000 people, which is 4.9% of the population of the entire basin.

The damage, the cost of the recommended protective structures and activities, their effectiveness were made with the Russian method of assessing the probability of damage from harmful effects of water and evaluation of the effectiveness of preventive measures of water, VIEMS, 2006.

The calculations showed that the total socio-economic cost of freshet at 1% occurrence on the Selenga River is 35552,364 mln. rub., including the territories and objects – 18464,087 mln. rub. and population – 1851,552 mln. rub.

It should be noted that only for the municipality of the city of Ulan-Ude, it can reach 27323,610 mln. rub.

In addition to the city, the hazardous area contains the following major industrial settlements and administrative centers: Naushki (329,46 mln. rub.), Kabansk (1342,61 mln. rub.), Ilyinka (944,56 mln. rub.), Sotnikovo (435.824 mln. rubles.), Nizhniy Sayantuy (415,509 mln. rub.), Poselye (1408,27mln. rub.) and others. The damage from the flood caused by ice dam is up to 1019,181 mln. rub. and as a result of erosion and collapse of the coastline for 25 years - 147577,392 thousand rub.

In the course of an integrated consideration of the type of settlement of its socio-economic development, total amount of damages, the level of emergency, the current state waterworks, recommendations to carry out certain activities on engineering protection from the negative effects of water were justified and proposed for 54 settlements. This requires the construction and reconstruction of dams of 172,104 km length, construction and repair of coast protection respectively – 33,971 km and dredging – 11,614 km, the total value of projected works: 5,804,247.36 thousand. rub. However, the cost-effectiveness calculations have shown that the usefulness of the activities is not economically beneficial for all settlements.

Thus, based on the research and the results of calculations of economic efficiency, the analysis of federal and republican target programs, in some cases, taking into account the actual situation, the Program of implementation of the offered activities was developed. In general, for all 26 territorial areas of settlements, an engineering protection from flooding is required, taking into account individual associations surrounding 3 villages near the city of Ulan-Ude and 9 villages of the Kabansky region. In 4 localities within which coast is exposed to intense erosion and collapse, the implementation of protective measures is economically feasible, but in 5 - is necessary.

Thus, the primary event proposed for 13 settlements, in the second place – for 7, and the third – for 11. In total costs 5,578,904.843 thousand rub. It is supposed that the probabilistic damage amounting to 36121,136 mln. rub. will be prevented, including prevention from flooding and congestion (36036,331 mln. rub.) and coast line collapse (84804,53 thousand rub.) and protecting of the population (for 38169 people).

In conclusion, it should be noted that within this research, the Selenga River is examined directly in detail. It is for certain, that non-exhaustive inclusion of other important tributaries cannot give a full picture of generalizing throughout the basin. That is why the complexity of the study was to consider the formation of floods in the basin, as the excluded tributaries provide large water supply and significantly affect the passage of floods on the Selenga. It is hoped that continuation of the research on proven methods will cover the whole Selenga basin, including its Mongolian part and other major rivers of the Lake Baikal basin.

References

Гармаев Е.Ж., Христофоров А.В. (2010): Водные ресурсы рек бассейна озера Байкал: основы их использования и охраны. Новосибирск: Академическое издательство «ГЕО». Publication in Russian language. [Garmaev E.Zh., Khristovorov, A.V. (2010): Water Resources of the Rivers of the Lake Baikal Basin: Basics of Their Use and Protection. Novosibirsk: Academic Press "Geo"].

Борисова Т.А. (2013): Природно-антропогенные риски в бассейне оз. Байкал. Новосибирск: Академическое издательство «ГЕО». – 126 с. Publication in Russian language. [Borisova T.A. (2013): Natural and man-made risks in the Lake Baikal Basin. Novosibirsk: Academic Press "Geo": 126].

Ресурсы поверхностных вод СССР. - Ленинград: Гидрометеоиздат, 1972. – Т.16. – Вып. 2. - 586 с.; 1973. – Т.16. – Вып.3. - 400 с. Publication in Russian language. [(1972): Surface water resources of the USSR. Leningrad: Gidrometeoizdat. - Т.16. - Vol. 2. - 586; 1973 - Т.16. - Vol.3.: 400].

Борисова Т.А., Бешенцев А.Н. (2011): Территориальная оценка риска от наводнений в Байкальском регионе в условиях экологических ограничений // Безопасность жизнедеятельности. Москва: Изд-во «Новые технологии». – № 12. – С.12–19. Publication in Russian language. [Borisova T.A., Beshentsev A.N. (2011): Territorial assessment of flood risk in the Baikal region in terms of environmental restrictions // Health and Safety. Moscow: Publishing House "New Technologies". - № 12: 12-19].

Challenges for Science-Based IWRM Implementation in Mongolia: Experiences from the Kharaa River Basin

Daniel Karthe[1,2], Sonja Heldt[1,3]

[1] Department Aquatic Ecosystem Analysis and Management, Helmholtz
Center for Environmental Research, Magdeburg, Germany.
Corresponding author: daniel.karthe@ufz.de

[2] Department of Geography, Georg-August University, Göttingen, Germany

[3] Department of Civil Engineering, Duisburg-Essen University, Germany

[4] Faculty of Environmental Sciences, Technical University, Dresden,
Germany

Abstract

Like its central Asian neighbors, highly continental Mongolia is not only a water-scarce but also a data-scarce country with regard to environmental information. At the same time, regional effects of global climate change, major land use changes, a booming mining sector and growing cities with insufficient and decaying water and wastewater infrastructures result in an increasingly unsustainable exploitation and contamination of ground and surface water resources putting at risk both aquatic ecosystems and human health.

For the mesoscale (15.000 km²) model region of the Kharaa River Basin (KRB), we investigated (1) the current state of aquatic ecosystems, water availability and quality; (2) past and expected future trends in these fields and their drivers; (3) water governance structures and their recent reforms; (4) technical and non-technical interventions as potential components of an IWRM. By now, the KRB is recognized as one of the most intensively studied river basins of the country, and considered a model region for science-based water resources management by the Mongolian government which recently adopted the IWRM concept in its National Water Program. Based on the scientific results and practical experiences from a six-years project in the KRB, the potentials and limitations of IWRM implementation in similarly data-scarce regions are discussed.

1. Introduction

IWRM, which Global Water Partnership defined as the "process which promotes the coordinated development and management of water, land and related resources, in order to maximize the resultant economic and social welfare in an equitable manner without compromising the sustainability of vital ecosystems (GWP-TAC, 2000) is considered as the most appropriate general concept for water management and more specifically, in developing and transformation countries (Heldt 2014; Snellen & Schrevel 2004). The United Nations have recommended IWRM as a straightforward approach for meeting the Millenium Development Goals in their World Water Development Report (2006) (Anderson et al. 2008). The concept of IWRM has gained wide acceptance in the majority of countries worldwide in the last 20 years, but the actual implementation of IWRM is lagging behind formal inclusion into policies, strategies and laws (Borchardt & Ibisch 2013).

The broadness of GWP's IWRM definition makes it a framework for normative and strategic water management that leaves space for many different types of approaches at the operational level adopted to the respective situation in specific countries or regions (Snellen & Schrevel 2004; Mitchell 2004). However, the concept has also been heavily criticized for vagueness of the formulation. Since the definition tries to be as holistic as possible, it incorporates both sustainable development and cross-sectorial planning simultaneously, which both are extreme complex and comprehensive for themselves (Jeffrey & Gearey 2006). Hering & Ingold (2012) have doubts on the overall feasibility and implementabilty of IWRM, which from their point of view is lacking methodology. Biswas (2008) considers IWRM an 'undefinable and unimplementable concept' that is politically preferable simply because of its vagueness that allows people to continue with what they had always done by only renaming their process after the current political correct terms.

The sufficient access to comprehensive and reliable data is a prerequisite for the transition from IWRM concepts to actual implementation. The support of all engaged parties with unbiased and trustworthy information can only be delivered by independent science (Stålnacke & Gooch 2010). However, poorer countries are often not able to operate independent research and monitoring due to lacking financial, infrastructural or educational capacities. Therefore they often have to rely on financial support from external donors and/or data collected in the context of international collaborations. While this brings a risk of bias, international scientific and development cooperation that strongly incorporates capacity building can be very effective (Anderson et al. 2008; Heldt 2014).

2. Scientific Basis for IWRM Planning in the Kharaa River Basin

Since 2004, the Mongolian government has gradually adopted the IWRM concept as the guiding principle for national water policy and designated 29 river basins of national importance (Karthe et al. 2014). From 2006 until 2013, an interdisciplinary research project (www.iwrm-momo.de) worked towards improvements in the scientific basis for an IWRM in Mongolia's Kharaa River Basin as a case study for the country and region (MoMo Consortium 2009; Karthe et al. 2011; Menzel et al. 2011; Karthe et al. 2014). For the previously data-scarce region, the project assessed the current state of water resources and the impacts of environmental and socio-economic changes on water availability and quality.

2.1. Water Availability and Its Determinants

The highly continental subarctic climate in the KRB is characterized by very cold winters with temperatures regularly reaching -40°C and below, warm summers and an average annual precipitation of around 300 mm. With regard to precipitation, there are large regional and interannual variations (Karthe et al. 2013; Menzel et al. 2011). Under present climate and land use conditions, on average more than 85% of the precipitation is lost due to evapotranspiration, leaving only a small portion of water available for stream flow and groundwater recharge. River runoff is linked to strong summer rainfall events and the melt of snow and river icings in spring, which cause in an inter-seasonal redistribution of water resources. Common subarctic phenomena in the catchment are snow, permafrost and aufeis (Hülsmann et al 2014).

Climate variability and change is one major driver of hydrological trends in drylands. In Mongolia, air temperature has increased by 1.8K since the 1940s (Batimaa 2006), and precipitation decreased in some parts of the country, including the Western slopes of the Khentii mountains, where the headwaters of the Kharaa river originate (Batimaa et al. 2008). For the 21st century, climate scenarios predict a further increase of mean annual air temperature between 2.6 and 5.1K in Mongolia, while mean annual precipitation is expected to increase by 20 to 86 mm (Karthe et al. 2013). This may, via increased evaporation, lead to a decrease in runoff and an increased likelihood of water scarcity (Batimaa et al. 2008). However, there are high uncertainties regarding future precipitation, and projected precipitation changes between different climate models for the same scenario are often larger than between two scenarios (Menzel et al. 2008). Some authors even predict a significant decrease in precipitation in northern Mongolia from the 2070s onwards (Sato et al. 2007).

Land use changes may be of an even larger hydrological relevance than climate change, having effects on both water quantities and quality (Karthe et al. 2014). Causes of land use change include the extension of agricultural activities, rising livestock numbers, logging and wildfires. A massive expansion and intensification of the agricultural sector is currently being propagated by national agropolitics (Priess et al.2011) and will lead to an increase in agricultural lands of 50 to 100%, with the side-effect of losses of large areas of natural vegetation and their related ecosystem services and functions (Schweitzer 2012). Since water is already the most important limiting factor for agricultural production, irrigation is seen as one solution to this dilemma, even more so because climate change leads to an increase in agricultural water demand in large parts of Mongolia, including parts of the Kharaa Basin (Menzel et al. 2008).

Between April 2000 and May 2012, a total of 200,000 ha, or about 14 %, of the forested area in the KRB have been affected by wildfires to various degrees. The majority of wildfires occurred in forested areas of the headwater regions, thereby altering almost pristine ecosystems that are major importance for regulating the water regime (Schweitzer 2012; Karthe et al. 2014). The comparison of precipitation, soil moisture and evapotranspiration on a steppe and shrubland site in the upper KRB revealed that there is only a negligible freshwater generation in sparsely vegetated steppe areas, whereas shrubland and taiga play an important role for water availability for the entire river basin (Minderlein 2014).

The increasing scarcity of water will probably trigger competition between different water-users. Of particular relevance in this context is the booming mining sector. In the KRB, mining sites cover around 130 km², but relatively little is known about their water consumption. Given the geological conditions and technology in use, a consumption of around 7 m³ of water per 1 m³ of ore seems to be realistic (MoMo Consortium 2009).

2.2. Water Quality and Aquatic Ecosystem Status

The Kharaa River Basin is, by international standards, characterized by a relatively low population density (about 8 people/km²). However, localized concentrations of population, an often poor state of urban waste water infrastructures, high livestock densities in the riverine floodplains and large-scale mining activities are potential threats to the aquatic ecosystems of the Kharaa. Among the key stressors affecting water quality and the aquatic ecosystem of the Kharaa are rising nutrient inputs, high fine sediment loads, the loss of habitat integrity with hyporheic and riparian zones and mining-related influxes of toxic substances (Karthe et al. 2014).

A state inventory for surface water in Mongolia conducted in 2003 showed that even though most rivers in the country were in relatively pristine condition, at least 23 rivers in 8 provinces were morphologically changed and/or polluted due to mining activities (Batsukh 2008), including the Kharaa-Orkhon-Selenga river system. Studies in the Kharaa and adjoining river basins show that gold mining is a major polluter (Thorslund et al. 2012; Chalov et al. 2012; Pfeiffer et al. 2014) and that it drastically affects the ecology of diatom, macrozoobenthos and fish communities (Krätz et al. 2010; Avlyush 2011). Moreover, high concentrations of nutrients, boron and a high electrical conductivity which were observed in several locations in the lower Kharaa River Basin are indications of groundwater contamination by improper waste water disposal (Hofmann et al. 2014).

In recent years, increasing nutrient loads could be observed for the Kharaa River, with a considerable longitudinal deterioration of water quality. Results of nutrient emission modeling confirm that urban settlements are the main sources for nitrogen and phosphorus emissions (Hofmann et al. 2011). Since only 35 % of the total population in the river basin is connected to WWTPs, unconnected urban areas represent an important proportion of the total emissions. At the same time, due to their poor state, WWTPs themselves are substantial point sources for nutrient influxes. Agriculture contributes 35% of total nitrogen and 32% of total phosphorus emissions (Hofmann et al. 2010), mostly through erosion from cultivated land and fallows. Sediment input caused by river bank erosion is a significant emission pathway for phosphorus (Hartwig et al. 2012; Theuring et al. 2013). This process is triggered by an increasing degradation of riparian vegetation due to high livestock densities with free access to the running waters.

Furthermore, high fine sediment loads constitute a major problem themselves since sediment-induced clogging inhibits essential habitat functions of the hyporheic interstitial (Schäffer 2008; Hartwig et al. 2012). As a consequence, functional shifts of the macroinvertebrate community and fish fauna have already been observed (Hofmann et al. 2011; Avlyush et al. 2013). Isotope-based sediment source fingerprinting techniques identified riverbank erosion (74.5%) and surface upland erosion (21.7%) as the main contributors to the suspended fine sediment load in the catchment (Hartwig et al. 2012, Theuring et al. 2013). Using erosion risk scenarios, Priess et al. (2014) show that erosion could increase more than twofold in the steppe regions of the lower KRB and up to sevenfold in the forested and mountainous upper KRB due to the combined impacts of land use and climate changes. However, the authors also argue that land management practices which are adapted to the semi-arid steppe environment (e.g., mulching of croplands with wheat straw) could in the future help to reduce soil erosion.

2.3. Water in Urban Mongolia: the Example of Darkhan

Integrated water resources management is particularly complex in the context of urban areas. Cities are not only characterized by a local concentration of water abstractions and waste water generation. They typically depend on water resources from their hinterland, and contaminate rivers further downstream. Therefore, sustainable urban water management needs to look beyond city borders (Anthonj et al. 2014). Moreover, an integrated concept for urban water management requires a clearly defined strategy for priority setting (Rost et al. 2014). This is particularly relevant in the context of transition countries like

Mongolia experiencing a rapid urbanization which often outpaces urban planning and development.

In case of Darkhan, the largest city in the Kharaa River Basin and third-largest city in Mongolia, key challenges in urban water management include the efficient provision of adequate quantities of safe drinking water, sanitation services and waste water management.

Provision of safe drinking water	Mining activities upstream of Darkhan city are potential sources of water contamination. In 2007, an accident in an illegal gold mining operation in Khongor Sum, just upstream of Darkhan, contaminated local drinking water sources with mercury and cyanides (Hofmann 2008). The ash basin of Darkhan's thermal power plant, which is located near two of the city's drinking water wells, is heavily contaminated with arsenic (Pfeiffer et al. 2014). Groundwater in several parts of Darkhan shows signs of wastewater contamination, e.g. through elevated boron and chloride levels (Hofmann et al. 2014). Shallow wells located in the proximity of latrines are potential sources of water-borne disease transmission, but reliable data on microbiological water quality and water-borne infections are almost non-existant (Karthe et al. 2012).
Water losses in centralized supply system	Large apartment blocks in Darkhan are connected to the city's centralized cold and warm water distribution system. Until repair works were started in 2012, leakage losses were about 50%. Moreover, low water fees and the absence of meters leads to a very high per capita consumption, which was around 400L/day in 2009 (MoMo Consortium 2009).
Limited access to water in peri-urban ger areas	Peri-urban ger areas receive their drinking either from pipe- or truck-fed water kiosks, by private wells or by surface water. The daily per capita consumption of water in Darkhan's ger areas averages about 12l, which falls below the minimum UNICEF and WHO recommendations of 15l to 25l. This may be explained both by a relatively high cost and the more difficult access to water as compared to tap water (Sigel et al. 2012).
Deficient urban waste water treatment	Since its commissioning in 1968, no constructional changes besides some minor works in 1978 and 1998 have been undertaken at Darkhan's WWTP (Heppeler 2012). The plant was not designed to eliminate nitrogen and phosphorous nor is there a disinfection stage. Oxygen depletion the risk of eutrophication are likely for the receiving water body, the Kharaa River. The expected growth of Darkhan's population, which is expected to double to more than 160,000 inhabitants by 2040, may aggravate this problem unless improvements in waste water treatment are made (Karthe et al. 2015).

Table 2: Urban Water Management Challenges in Darkhan

The situation in Darkhan is typical for that of major cities in Mongolia, such as most Aimag capitals. In most cases, a transparent priorization of measures is difficult due to a lack of data about both environmental contamination and urban water consumption.

3. Challenges for IWRM Implementation in Mongolia

There are currently several challenges for the implementation of an IWRM in Mongolia, including an ongoing transformation of the institutional framework, a lack of financial capacities and qualified staff in the water sector, and a general shortage of environmental data, which hampers the identification of priority measures and the monitoring of progress.

Since the 1990s, Mongolia has undergone a far-reaching political and socioeconomic transformation processes which has had a strong impact on water management (Houdret et al. 2013). The Mongolian water sector had been dominated by central planning during Soviet times, but the water administration was almost entirely abolished during the transformation in 1990. New institutions were gradually established, and under its 2004 water law, Mongolia decided to introduce water governance at the river basin level. To this end, 29 river basins were officially identified by the then Ministry of Nature, Environment and Tourism (MNET) on the basis of hydrological criteria but also considering economic and political implications. At the national level, two new institutions were created to implement IWRM: the Water Authority as the main agency to implement RBM and the National Water Council (NWC), assembling representatives of seven different ministries in order to enhance a cross-sectoral approach to water governance (Karthe et al. 2014). However, the implementation of these institutional changes faced severe challenges both at national and at river basin levels. Deficiencies in horizontal interplay, i.e. coordination and cooperation between institutions of the different sectors (such as, for instance, construction, mining, environment, agriculture) and concerning the vertical interplay, i.e. cooperation between institutions of the different political levels (communities, provinces, national level) were major obstacles to effective water governance. Another key difficulty for the implementation of RBM was the 'problem of fit', i.e. the mismatch between the boundaries of the newly created river basins and the ones of the existing public administrations, such as the provincial parliament, the local council, or the provincial environment department (Horlemann & Dombrowsky 2012). Even though several River Basin

Councils were created, their scope of action has been seriously restricted by legal power, financial resources and political mandate. A key issue of concern in this context was the unavailability of adequate funding for the RBCs (only the chairperson and the secretary were paid, no budget for the functioning of the council). As a result, only in the basins that were supported by external donors could RBCs de facto be created and partially started developing RBMPs (Livingstone et al. 2009; Lkhagvadorj 2010; Karthe et al. 2014).

As a response to the persisting difficulties in the implementation of IWRM and in the face of increasing problems of water pollution and growing water demand, the Mongolian government adopted a new water law in 2012. At the national level, environmental and water governance gained stronger political influence, including the Ministry of Environment and Green Development (MEGD, i.e. the former MNET) and the National Water Council, which were also placed under the auspices of the prime minister. A new entity for the financing, construction and maintenance of water infrastructure, Mongol Us, was created. Moreover, the new budget law provides for significant financial means for local governments, potentially benefitting local environmental governance and the implementation of RBMPs (Lkhagvadorj 2012; Karthe et al. 2014). At the river basin level, RBCs were dissolved and river basin administrations (RBAs) formed instead, with plans to revitalize RBCs in the future.

In large parts of Mongolia, the scarcity of reliable environmental data is a considerable obstacle for river basin management planning. Due to a very limited monitoring network, hydrometeorological data and information on local and regional surface and groundwater availability and quality are often lacking. Moreover, several environmental monitoring stations and programs have at least temporally ceased to operate since the 1990s, and relatively little is known about the current ecological state of many surface water bodies. With regard to environmental information, the Kharaa River Basin is an exception, however. For this reason, it is now by many, including relevant ministries and the National Water Committee, considered as a model region with reference data sets and methodologies for IWRM development in northern Mongolia.

4. Conclusions

In recent years, Mongolia has experienced a growing water demand, a wide-spread deterioration in surface and ground water quality, and a degradation of aquatic ecosystems. This is caused by a multitude of factors, including climate change, mining-driven economic growth and rapid urbanization. Since water availability is limited for climatic reasons, any overexploitation of water re-sources is likely to result in a further local or regional increase of water scarcity and a growing competition between different users. In such a context, IWRM seems a promising approach since it considers ecological and socioeconomic development in a holistic way; and science-based IWRM would ideally result in a rational, unbiased basis for decision-making. This is fully acknowledged by the Mongolian government. However, limited institutional, legal, financial and staff capacities still inhibit the implementation of a national IWRM program. These challenges, combined with a poor availability of un-biased data currently lead to a dependence on the collaboration with foreign donors and research projects, which can nevertheless be beneficial if it enables local partners for a self-sustained implementation in the future.

Acknowledgements

The results presented in this paper are based on the research and develop-ment project „Integrated Water Resources Management in Central Asia: Model Region Mongolia", funded by the German Federal Ministry of Education and Research (BMBF) in the framework of the FONA (Research for Sustainable Development) initiative (grant no. 033L003). We acknowledge the support pro-vided by the Project Administration Jülich (PTJ) and the BMBF/International Bureau in the context of the "Assistance for Implementation" (AIM) scheme.

References

Anderson A., Karar E. and Farolfi S. (2008) Synthesis: IWRM lessons for im-plementation. Water SA 34(6): 665-669.

Anthonj, C.; Beskow, S.; Dornelles, F.; Fushita, A.T.; Galharte, C.A.; Galvão, P.; Gatti Junior, P.; Gücker, B.; Hildebrandt, A.; Karthe, D.; Knillmann, S.;

Kotsila, P.; Krauze, K.; Kledson Leal Silva, A.; Lehmann, P.; Moura, P.; Periotto, N.A.; Rodrigues Filho, J.L.; Lopes dos Santos, D.R.; Selge, F.; Silva, T.; Soares, R.M.; Strohbach, M.; Suhogusoff, A.; Wahnfried, I.; Zandonà, E.; Zasada, I. (2014): Water in Urban Regions: Building Future Knowledge to Integrate Land Use, Ecosystem Services and Human Health. Halle/Saale, Rio de Janeiro, Berlin: German National Academy of Sciences Leopoldina, Brazilian Academy of Sciences, German Young Academy. 32 pages.

Avlyush S (2011) Effects of surface gold mining on macroinvertebrate communities. A case study in river systems in the North-East of Mongolia. Saarbrücken: Lambert Academic Publishing.

Avlyush S., Schäffer M. and Borchardt D. (2013): Life cycles and habitat selection of two sympatric mayflies under extreme continental climate (River Kharaa, Mongolia). International Review of Hydrobiology 98 (3):141 – 154.

Batimaa P. (2006): Climate Change Vulnerability and Adaptation in the Livestock Sector of Mongolia. A Final Report Submitted to Assessments of Impacts and Adaptations to Climate Change (AIACC). Project No. AS 06, Washington.

Batimaa P, Batnasan N. and Bolormaa B. (2008): Climate change and water resources in Mongolia. In: Basandorj, B. and D. Oyunbatar (Ed.) 2008: International Conference "Uncertainties in water resource management: causes, technologies and consequences", pp. 7-12. Jakarta: IHP Technical Documents in Hydrology No. 1.

Batsukh N; Dorjsuren D, Batsaikhan G (2008) The water resources, use and conservation in Mongolia. First national report. Ulaanbaatar: National Water Committee.

Biswas, A.K. (2008): Integrated Water Resource Management: Is It Working? International Journal of Water Resources Development 24(1):5–22.

Borchardt D & Ibisch R (Eds. 2013): Integrated water resources management in a changing world: lessons learnt and innovative perspectives. London: IWA Publishing.

Chalov S.R., Zavadsky A.S., Belozerova E.V., Bulacheva M.P., Jarsjö J., Thorslund J. and Yamkhin J. (2012): Suspended and Dissolved Matter Fluxes in the Upper Selenga River Basin: Synthesis. Geography, Environment, Sustainability 02(05):78-94.

Global Water Partnership Technical Advisory Committee (=GWP-TAC) (2000): Integrated Water Resources Management. Stockholm, Sweden.

Hartwig M., Theuring P., Rode M. and Borchardt D. (2012): Suspended sediments in the Kharaa River catchment (Mongolia) and its impact on hyporheic zone functions. Environmental Earth Sciences 65(5):1535-1546.

Heldt S. (2014): The EU-WFD as an Implementation Tool for IWRM in non-European Countries. Case Study: Drafting a River Basin Management Plan for the Kharaa River in Northern Mongolia. Master thesis, Duisburg-Essen University, Germany.

Hering J. and Ingold K.M. (2012): Water Resources Management: What should be integrated? Science 336 (6086):1234-1235.

Hofmann J (2008) Bericht über die Untersuchungen von Grundwasser und Boden auf Schwermetalle und Cyanid in Khongor Sum. doi: 10.13140/2.1.4306.5928

Hofmann J., Hürdler J., Ibisch R., Schaeffer M. and Borchardt D. (2011): Analysis of Recent Nutrient Emission Pathways, Resulting Surface Water Quality and Ecological Impacts under Extreme Continental Climate: The Kharaa River Basin (Mongolia). International Review of Hydrobiology 96(5): 484-519.

Hofmann J., Venohr M., Behrendt H., Opitz D. (2010): Integrated water resources management in central Asia: nutrient and heavy metal emissions and their relevance for the Kharaa River Basin, Mongolia. Water Science and Technology 62(2):353-363.

Hofmann J., Watson V. and Scharaw B. (2014): Groundwater quality under stress: contaminants in the Kharaa River basin (Mongolia). Environmental Earth Sciences. doi: 10.1007/s12665-014-3148-2

Horlemann L, Dombrowsky I (2012) Institutionalising IWRM in developing and transition countries: the case of Mongolia. Environmental Earth Sciences 65(5):1547-1559.

Houdret A., Dombrowsky I. and Horlemann L. (2013): The institutionalization of river basin management as politics of scale - insights from Mongolia. Journal of Hydrology. http://dx.doi.org/10.1016/j.jhydrol.2013.11.037

Hülsmann L., Geyer T., Schweitzer C., Priess J. and Karthe D. (2014): Initial Results and Limitations of the SWAT Model applied to the Kharaa River Catchment in Northern Mongolia. Environmental Earth Sciences. doi: 10.1007/s12665-014-3173-1.

Jeffrey P. and Gearey M. (2006) Integrated water resource management: lost in the road from ambition to realization? Water Science & Technology, 53(1):1-8.

Karthe D., Borchardt D and Kaus A. (2011): Towards an Integrated Water Resources Management for the Kharaa Catchment, Mongolia. In: Гуринович, А.Д. (Ed.) (2011): Proceedings of the IWA 1st Central Asian Regional Young and Senior Water Professionals Conference, Almaty/Kazakhstan, pp. 79-93.

Karthe D., Heldt S., Houdret A. and Borchardt D. (2014): IWRM in a country under rapid transition: lessons learnt from the Kharaa River Basin, Mongolia. Environmental Earth Sciences. doi:10.1007/s12665-014-3435-y

Karthe D., Heldt S., Rost G., Londong J., Ilian J., Heppeler J., Khurelbaatar G., Sullivan C., van Afferden M., Stäudel J., Scharaw B., Westerhoff T., Dietze S., Sigel K., Hofmann J., Watson V. and Borchardt, D. (2015): Modular Concept for Municipal Water Management in the Kharaa River

Basin, Mongolia. In: Borchardt, D.; Bogardi, J. & Ibisch, R. (2015): Integrated Water Resources Management: Concept, Research and Implementation. In press.

Karthe D, Sigel K, Scharaw B, Stäudel J, Hufert F and Borchardt D (2012): Towards an integrated concept for monitoring and improvements in water supply, sanitation and hygiene (WASH) in urban Mongolia. Water & Risk 20:1-5.

Krätz D., Ibisch R., Saulyegul A., Ganganmurun E., Sonikhishig N. and Borchardt D. (2010): Impacts of Open Placer Gold Mining on Aquatic Communities in Rivers of the Khentii Mountains, North-East Mongolia. Mongolian Journal of Biological Sciences 8(1):41-50.

Livingstone A., Erdenechimeg C. and Oyunsuvd A. (2009): Needs Assessment on Institutional Capacity for Water Governance in Mongolia. United Nations Development Program (UNDP) and Government of Mongolia. Ulaanbaatar.

Lkhagvadorj A. (2010): Fiscal federalism and decentralization in Mongolia. Potsdam: Universitätsverlag Potsdam.

Lkhagvadorj A. (2012): Analysis on Mongolia's Integrated Budget Law. Ulaanbaatar: SDC Publications.

Menzel L., Hofmann J. and Ibisch R. (2011): Untersuchung von Wasser- und Stoffflüssen als Grundlage für ein Integriertes Wasserressourcen – Management im Kharaa-Einzugsgebiet (Mongolei). Hydrologie und Wasserbewirtschaftung 55(2):88-103.

Menzel L., Aus der Beek, T., Törnros T. et al. (2008): Hydrological impact of climate and landuse change – results from the MoMo project. In: Basandorj, B. and D. Oyunbatar (Ed.) (2008): International Conference "Uncertainties in water resource management: causes, technologies and consequences", pp. 13-18. Jakarta: IHP Technical Documents in Hydrology No. 1.

Minderlein S. (2014): Evapotranspiration and energy balance dynamics of a semi-arid mountainous steppe and shrubland site in northern Mongolia. Environmental Earth Sciences. doi: 10.1007/s12665-014-3335-1

Mitchell B. (2004): Comments by Bruce Mitchell on Water Forum contribution "Integrated Water Resources Management: A Re-assessment by Asit K. Biswas". Water International 29(3):398-399.

MoMo Consortium (2009): Integrated Water Resources Management for Central Asia: Model Region Mongolia (MoMo). Case Study in the Kharaa River Basin. Final Project Report.

Pfeiffer M., Batbayar G., Hofmann J., Siegfried K., Karthe D. and Hahn-Tomer S. (2014): Investigating arsenic (As) occurrence and sources in ground, surface, waste and drinking water in northern Mongolia. Environ Earth Sciences. doi: 10.1007/s12665-013-3029-0

Priess J.A., Schweitzer C., Batkhishig O., Koschitzki T. and Wurbs D. (2014): Impacts of agricultural land-use dynamics on erosion risks and options for land and water management in Northern Mongolia. Environmental Earth Sciences. doi: 10.1007/s12665-014-3380-9

Sato T., Kimura F. and Kitoh A. (2007): Projection of global warming onto regional precipitation over Mongolia using a regional climate model. Journal of Hydrology 333(1):144-154.

Schäffer M., Ibisch R.B. and Borchardt D. (2008): Invertebrate Lebensgemeinschaften als Indikatoren für Landnutzungseffekte im Norden der Mongolei. In: Deutsche Gesellschaft für Limnologie [=DGL] (2009): Erweiterte Zusammenfassungen der Jahrestagung 2008 (Konstanz), Hardegsen 2009, pp. 308-312. Krefeld: Eigenverlag der DGL.

Schweitzer C. (2012): Modelling land-use and land-cover change and related environmental impacts in Northern Mongolia. Doctoral dissertation at Martin-Luther University Halle-Wittenberg, Germany.

Sigel K, Altantuul K & Basandorj D (2012) Household needs and demand for improved water supply and sanitation in peri-urban ger areas: The case of Dar-khan, Mongolia. Environ Earth Sci 65(5):1561-1566. doi: 10.1007/s12665-011-1221-7

Snellen W.B. and Schrevel A. (2004) IWRM: for sustainable use of water 50 years of international experience with the concept of integrated water management. Wageningen, the Netherlands: Ministry of Agriculture, Nature and Food Quality.

Stålnacke P. and Gooch G.D. (2010): Integrated Water Resource Management. Irrigation and Drainage Systems 24(3-4):155–159.

Theuring P., Rode M., Behrens S., Kirchner G., Jha A. (2013): Identification of fluvial sediment sources in a meso-scale catchment, Northern Mongolia. Hydrological Processes 27(6):845-856.

Thorslund J., Jarsjö J., Chalov S.R. and Belozerova E.V. (2012): Gold mining impact on riverine heavy metal transport in a sparsely monitored region: the upper Lake Baikal Basin case. Journal of Environmental Monitoring 14(10):2780-92.

The EU-WFD as an Implementation Tool for IWRM in non-European countries – Case Study: Mongolia

Sonja Heldt[1], Daniel Karthe[2], Christian Feld[3]

[1] Department of Civil Engineering, Duisburg-Essen University, Germany
Corresponding Author. Email: sonja.heldt@uni-due.de
[2] Department Aquatic Ecosystem Analysis and Management, Helmholtz
Center for Environmental Research, Magdeburg, Germany
[3] Department Aquatic Ecology, Duisburg-Essen University, Germany

Abstract

Globally and especially in the naturally water-scarce Mongolia, the quantity but also the quality of water resources are threatened by a rising pressure caused by population and economic growth and the resulting environmental alterations. The IWRM concept as well as the EU-WFD aim at the protection and restoration of water resources and aquatic ecosystems byapplying integrative management strategies. Whereas the IWRM concept is criticized mostly because of its vagueness that misses clear implementation guidance, the EU-WFD represents a clearly defined framework to achieve a sustainable management of natural water resources. Despite being a specifically European framework, the diverse preconditions in the EU member states require the EU-WFD to be implementable in very different country-specific contexts. In this way, the EU-WFD may potentially also be an appropriate tool to enhance sustainable water management in non-European countries that actually have problems in the implementation of their national IWRM concepts.

Introduction

Water is an essential resource for all types of life, but a sufficient availability or quality cannot be taken for granted in all parts of the world. Besides its vital importance for humans, water is also a necessary prerequisite for economic and social development. The drastic growth of the world's population and economy since the start of the industrial revolution, and the along-going increasing demand for fresh water, are inevitably linked with serious alterations of the natural water cycle, water quality and aquatic ecosystems. Therefore, a sustainable management of the existing water resources and its appending environment is essential to ensure sufficient water supply in future.

Globally, the instancy of water problems and the resulting need for a comprehensive protection of water resources was recognized and discussed for the first time at the United Nation Water Conference in Mar del Plata, Argentina in 1977, where first principles for a totally new approach of sustainable water management were adopted. Later, the Earth Summit in Rio (1992) recommended the new concept of Integrated Water Resource Management (IWRM) and put it on top of the international Agenda. In 2000, Global Water Partnership (GWP) published the first comprehensive definition of IWRM (GWP 2009; Snellen & Schrevel 2004). First implementation efforts uncovered the high complexity of the concept, as the success of IWRM depends on the coordination and integration of numerous stakeholders and institutions with partially different political and economic backgrounds and interests (Biswas 2004; Grooch & Stalnacke 2010; Snellen & Schrevel 2004).

In fact, the global emergence of IWRM had been preceded by similar considerations in Europe. During the 1970s, the European Community (EC) updated its environmental laws and thereby placed a high priority on the conservation and sustainable management of water resources (Rahaman et al. 2004). The process culminated in the adoption of the European Water Framework Directive (EU-WFD; EU legislation no. 2000/60/EC) in 2000. In order to reach its main objective i.e. to achieve and/or maintain a 'good ecological status' for all European water bodies, the directive promotes the integration of cross-sectional management as well as the participation of public and private stakeholders. Since its adoption the EU-WFD was supplemented by minor directives that contained regulations regarding groundwater (2006), floods (2007) and environmental quality standards (2008).

In Mongolia, recent developments in the water sector induced political reforms that made the idea of science-based River Basin Management (RBM) the core of Mongolia's national water policies and thus opened a window for the establishment of IWRM (Karthe et al. 2014c). Mongolia's national IWRM policy calls for basin-specific, sustainable water management that incorporates the adaptation to changing climate conditions and the resulting uncertainties in future water resources availability (Mongolian Ministry for Environment and Green Development (MEGD) 2012). However, the enforcement of this political and institutional effort has not been successful so far.

1. Water Management in Europe – an international model?

1.1. The European Water Sector

The European Union currently consists of 28 member states and covers a territory of 4,381,324 km^2. Therefore water management in Europe faces a multiplicity of environmental, political, socio-economic and cultural preconditions. Directives, that have to be enforced in all member states, such as the EU-WFD, are thus required to be highly flexible and applicable to a very broad range of country specific contexts.

Finland for example features boreal climate with warm summers and freezing winters, a population density of approximately 16 people per square kilometre with a minimum of 2 people per square kilometre in Lapland, thousands of lakes and rivers that cover more than 10 % of the entire territory and a nominal GDP of 35,600 € that is above European average (EC 2012a). Opposing to this, Malta is characterized by subtropical-mediterranean climate with very mild winters and warm to hot summers, a population density of 1,306 peoples per square kilometre (95 % of its entire territory are urban), no inland surface waters at all (EC 2012b) and a GDP below European average of 17,000 €.

The main concerns for national water management differ significantly within the European Union. In the Netherlands, for example, rural and urban citizens have virtually 100 % access to safe drinking water supply and sanitation with a sophisticated waste water treatment system. The main challenges for water management arise from the unnatural hydromorphology of heavily modified water bodies, the ecological degradation caused by pollution, partly originating from upstream areas in other countries, as well as from the building and maintenance of flood protection measures (EC 2012c). Contrasting to this, in Romania the status of local water resources is mainly threatened by pollution from rural and urban settlements without any or with insufficient waste water

treatment facilities and agricultural farms that are lacking proper storage of animal wastes (EC 2012d). In this way, the Dutch and the Romanian water management has to set totally different priorities in order to enable the implementation of the environmental objectives of the EU-WFD.

But even if the socio-economic and environmental preconditions seem to
be comparable (for example in France and in Germany), the political commit-
ment to implement the EU-WFD requirements can be substantially different.
France has adopted water management on river basin level in 1964 and, with
its six self-financing water agencies, has developed an advanced structure of
policies in this field, which was further adjusted to the requirements of the EU-
WFD with the Water Law of 2006 (GWP 2009; International Office for Water
2009). In Germany, RBM was no topic until the implementation of the EU-WFD.
Since 2003 the federal states are obliged to organize themselves in river basin
associations ("Flussgebietsgemeinschaften") that coordinate the implementation of the WFD for the respective river basins. The river basin associations are
subordinated and financed by the involved federal states and cannot be seen
as self-contained. The federal states set up administrative agreements for periods of five years (Flussgebietsgemeinschaft Elbe 2009).

1.2. The EU-WFD as a guideline for IWRM implementation in non- Euro-
pean countries?

While the IWRM concept and its basic definition have been criticized for being
too complex without providing a clear implementation guideline (Biswas 2004),
the EU-WFD is reputed as a stricter guideline providing clear definitions of its
principles, objectives and requirements. Both the IWRM concept and the EU-
WFD share the same overall objective, i.e. sustainable management of water
resources that assures sufficient water of good quality to meet future water
demands and preserve natural ecosystems.

Rahaman et al. (2004) conducted a systematical comparison of the
IWRM approach and the EU-WFD and thereby presented seven essential mismatches concerning the topics of gender awareness, integration between different sectors, decentralisation, participation of stakeholders, the focus on poverty, human-oriented management and the development of responsibilities at
the lowest appropriate level. While it may be argued that these points are of
little relevance for water resources management in the EU (many of them being

covered by other directives), this looks different when the EU-WFD alone is used as a blueprint for IWRM implementation in developing and transformation countries. According to the analyses by Rahaman et al. (2004), which compare IWRM and the EU-WFD as stand-alone approaches, the overlaps seem to be surprisingly small. However, the question is whether this is an adequate comparison that incorporates the intention behind both concepts. In fact, the purpose of the two already differs fundamentally. While the IWRM approach was developed as a general concept to address a global water crisis at a highly international level, the EU-WFD was adopted as a legal framework to unify the highly fragmented European environmental policy sector (Grambow 2013). Furthermore, IWRM claims to be a stand-alone approach that comprehensively manages all issues related to water in a cross-sectional way. In contrast the EU-WFD is one specific directive aiming at the protection of European water resources and water-related ecosystems. Consequently these formal differences also entail totally different requirements concerning focal points in principles and objectives (Heldt 2014).

A main aspect in Rahaman et al (2004)'s criticism of the EU-WFD is that it fails to connect aspects of nature protection with human needs. In general, the EU-WFD provides a strict guideline laying the foundation for a combined approach that relies on emission limits and quality standards to achieve a good status for all water bodies in Europe. Within this framework a guideline for social management aspects is not explicitly incorporated, but the environmental objectives required by the EU-WFD can basically not be met by excluding cross-sectional solutions. Achieving the required environmental objectives is closely connected to a comprehensive analysis of point and diffuse pollution sources that are related to anthropogenic activities within river basins. The focus on the improvement of water resource quality is closely linked to drinking water protection, while environmental protection and restoration also promote the attractiveness of surface waters for leisure and tourism. Moreover, inadequate sanitation and waste water treatment facilities have been determined as main source for water quality alterations in several EU Member States (WECF 2008; WHO 2010). As a consequence, the EU-WFD guidelines require the implementation of necessary measures to eliminate these sources, which in this case means the establishment of an appropriate waste water management system that besides water quality also increases the population's living and health standards. In this way the implementation of the EU-WFD guidelines will indirectly also gives incentives for the socio- economic development in the region.

As mentioned above, according to Rahaman et al. (2004) the EU-WFD does not understand the purpose of water as an economic good in the right way. But in fact the economic role of water is present in the directive from its first sentence (Preamble (1) EU-WFD). The enhancement of the ecological status is seen as an investment in future economic benefits (Preamble (17) EU-WFD) and economic analyses are required to support the introduction of economic instruments that set incentives for a responsible water use, like adjusted cost recovery systems (e.g. polluter-pays- principle, adequate water pricing etc.). Further, although decentralization is not obligatory, the directive recommends that 'decisions should be taken as close as possible to the locations where water is affected or used' (Preamble (13) EU-WFD). The constant dialogue between different policy sectors is also explicitly mentioned as a necessity (Preamble (16) EU-WFD).

Gender awareness, however, is neither directly nor indirectly addressed by the directive, but separately empowered by the EU Gender Equality Legislation that is composed of several directives (EC 2008). The adoption is thus a fixed part of the EU Community *acquis* (agreements) and obligatory for all EU Member States.

Similar to the Gender Equality Directive various other directives contribute to a framework that legally supports comprehensive integrative water management in the EU. Examples include the Habitats Directive (85/337/EEC), the Birds Directive (2009/147/EC), the Urban Waste Water Treatment Directive (91/271/EEC) and the Drinking Water Directive (98/83/EC).

For EU-Member States, supported by the supplemental set of EU legislation, the EU-WFD fulfils its task as a comprehensive and integrative water resource management approach that incorporates environmental issues, economic tools and the increased involvement of the public. In non-European contexts, however, the EU-WFD alone may not be suitable universal implementation tool for IWRM.

In two cases, the EU-WFD has been adapted by countries that are not members of the EU so far: (1) in case of candidate countries like Turkey (Sumer & Muluk 2011; ORSAM 2011) or (2) if the respective country is adjoining to a trans-national river basin that is partly located on EU territory, for instance in the Danube Basin (Bosnia-Herzegovina, Moldavia, Montenegro, Serbia and the Ukraine) (www.icpdr.org). But in these examples all countries are in close relation to the EU and benefit from its comprehensive support. In other cases, the immense investment costs connected to the implementation of the requirements of the EU-WFD constitute a significant barrier in transferring the directive per se, as other non-EU countries cannot rely on external funding or capacity support.

In general, the overall success of transferring the EU-WFD to countries that are not part of or directly supported by the EU will be highly dependent on the country-specific policy agenda. In particular, priorities tend to be different in low-income countries which often have to focus on basic human needs before environmental sustainability. Thus an implementation of the EU-WFD would not be a suitable solution to these countries water-related issues.

2. The EU-WFD as an implementation guideline for IWRM in Mongolia

2.1. The Mongolian Water Sector

The socio-economic preconditions in Mongolia can generally not be compared to the situation in developed countries of Western Europe. While in 2013 Germany for example had an HDI of 0.911 (UNDP 2014) and a GDP of 3.635 trillion USD (http://www.worldbank.org/en/country/germany), Mongolia had an HDI of 0.698 (UNDP 2014) and a GDP of 11.52 billion USD (http://www.worldbank.org/en/country/mongolia). Moreover, the most relevant anthropogenic pressures on water resources and aquatic ecosystems differ substantially between Mongolia and the Western European countries in particular. However, the range of environmental, socioeconomic and cultural settings in the EU is wide and increased with the east- and southeastwards expansions in 2004, 2007 and 2013. In a few regards, the new member states resemble Mongolia more closely than the Western European nations. The abrupt end of socialism caused a fundamental political and socioeconomic transition which brought benefits such as democratization but also challenges related to institutional reforms and the shift to free market economies. Nevertheless, in a few respects there are additional similarities between Mongolia and other regions of the EU (see table 1). Whenever water-related challenges and their causes are comparable, it seems reasonable to transfer established strategies and methods from Europe to Mongolia.

In contrast to the EU though, anthropogenic pressures in Mongolia often act as single and distinct stressors. For example, the identification of pollution sources and their impact on the environment is less complex, thus simplifying the derivation of appropriate countermeasures as compared to European multistressor settings.

	Mongolia	Corresponding situation in Europe
Population	Average pop. density: 2.3 people/km^2; Max. pop. density: 280 people/km^2 (Ulaanbaatar)	Average pop. density : 116 people/km^2; Min. pop. density: 2 people/km^2 (Lapland)
Urban Water Management	Population in the city centres is connected to obsolete central water supply and sanitation systems; population in peripheral areas has no safe water supply and access to sanitation (approximately half of Mongolia's population) (Genté 2013; Karthe et al. 2014b)	In Romania less approximately 20% of the urban population has no house connection to safe drinking water sources. In some urban areas in eastern Europe the percentage of households that have access to safe sanitation is only 60% (WHO 2010)
Rural Water management	42.5 % of Mongolia's population lives in rural areas (MEGD 2012) where safe drinking water supply and waste water disposal systems are typically not available or very restricted. Only 25% have access to improved sanitation and 39.2% to save drinking water (UNDP & UNICEF 2009)	23 % of the EU's population lives in rural areas; in parts of the Eastern EU (especially Romania), less than 20 % of the rural population have access to safe drinking water or sanitation (WHO 2010)
Surface and Groundwater Quality	Water quality degradations are found near/downstream of urban areas (nutrients), industrial and mining areas (heavy metals) and in rural areas characterized by high lifestock densities (Hartwig et al. 2012; Hofmann et al. 2014; Karthe et al. 2014a; Pfeiffer et al. 2014). Microbiological contamination of drinking water is a potential public health concern (Karthe et al. 2012)	Surface water quality in major rivers has improved over the past few decades; currently, the highest portion of surface water bodies with a poor chemical status can be found in the north of France and Sweden (www.eea.europa.eu)

State of Aquatic Eco-systems	Generally, most surface water bodies have a good ecological status, albeit with threats resulting from water quality impairments and a desiccation trend in some regions (Karthe et al. 2014a) as well as increasing pressures resulting from fishing (Kaus et al. 2011)	On average less than half of the European water bodies have a 'good ecological status', with a much better picture for Scandinavia (www.eea.europa.eu)
Main Anthropogenic Pressures	<ul><li>point pollution of surface water bodies from insufficiently treated urban and industrial waste water (Hofmann et al. 2011; Karthe et al. 2014b);</li><li>pollution of ground and surface water from latrines and inappropriate dump sites (Karthe et al. 2012; Uddin et al. 2014; Karthe et al. 2014b);</li><li>significant water abstractions for irrigation and mining activities (Hofmann et al. 2010)</li><li>influx of sediments and toxic substances due to mining operations (Hofmann et al. 2010; Thorslund et al. 2012)</li><li>river bank erosion, fine sediment and diffuse nutrient/pathogen release due to extensive livestock farming along river banks (Hartwig et al. 2012; Theuring et al. 2013)</li><li>Land use change, and in particular forest losses, with profound impacts on regional hydrological cycles (Minderlein & Menzel 2014; Karthe et al. 2014a)</li></ul>	Point pollution from insufficiently treated waste water (e.g. Romania, Bulgaria or Spain); Diffuse nutrient emission by settlements and agriculture (especially livestock farming) (e.g. Italy, Spain or Estonia); Chemical (especially heavy metals) and ecological degradation due to intensive mining (e.g. Poland, Slovakia or Hungary) (http://ec.europa.eu/)
Water governance framework	Since 2004, RBM and IWRM were progressively embraced as national strategies for water resources management, while water resources management was declared a priority field for government action; substantial institutional reforms, unclear/overlapping responsibilities and a lack of budget and capacities have so far impeded the implementation of IWRM (Horlemann & Dombrowsky 2012; Karthe et al. 2012)	Since 2000 the European Water Framework Directive represents a binding legislation for sustainable water management in Europe.

Table 3: Water-related challenges in Mongolia and the corresponding situation in Europe

2.2. IWRM in Mongolia

Similar to the social, cultural and economic sectors, Mongolia's political development was strongly affected by the socialist regime and its breakdown in 1990 (Houdret et al. 2013; Schweitzer 2012). Even though Mongolia has established a democratic multi-party system (Janzen 2012), political structures are still dynamic and the transition into a market economy ongoing. This leads to frequent changes in regulations, institutions and stakeholder interests.

The national constitution, the Water Law and the Environment Protection Law form the legal basis for water governance in Mongolia (Houdret et al. 2013; Karthe et al. 2014a). These legal regulations declare the water resources of Mongolia as a state property that is decoupled from land ownership. Hence, the national and regional governments are in charge of controlling water use and management. The National Water Program is a central strategy document aiming at the practical implementation of water management. Subsidiary regional development programs regulate the protection and rehabilitation of water resources as well as on the provision of sustainable access to safe drinking water and sanitation (MEGD 2012).

Since 1999, various national programs have been developed to implement the IWRM concept on a national level. Following the recognition of the Millennium Development Goals (2000) as a strategy for national development, the government declared the Mongolian Action Program for 21st century (2008) and in 2010 updated the National Water Program that was originally drafted in 1999 (Sigel 2012). Decentralization had been a key feature of the ongoing reform process in Mongolia and shifted the former strict top-down governance to a multi-level system. In the context of water governance, this led to the promotion of management at the river basin scale. Practically, 29 river basins of national importance were declared, for which river basin organizations began to be formed. However, the actual transfer to practice has been only weak so far (Houdret et al. 2013). Main reasons for deficits in the implementation process are a lack of financial, institutional and staff capacities and a serious shortage of data to support decision-making processes (Karthe et al. 2014a).

Several foreign actors have in the recent past promoted the implementation of IWRM and RBM in Mongolia by either conducting research projects or providing financial support. However, there is still no universally accepted agreement regarding the priorities for IWRM implementation (Houdret et al. 2013). This was recognized by the Mongolian government which is currently working towards a more clearly defined national IWRM concept, which provides a common basis for river basin management planning.

2.3. IWRM policy in Mongolia and the EU-WFD – potential synergies

A holistic consideration of the four Dublin Principles constitutes the basic philosophy of IWRM. Tables 2 to 5 describe the relation between this basic IWRM framework, the EU-WFD and current IWRM legislation in Mongolia (Heldt 2014).

Water should be treated as a finite and vulnerable resource that is essential to sustain life, development and the environment.	
EU-WFD:	**Mongolian IWRM policy:**
The EU-WFD partly refers to this principle as it aims to maintain water resources with a good status and prevent further deterioration of surface and groundwater resources. In this way the EU-WFD attempts to ensure the sufficient provision of good quality water to meet the future demand of its population. A main objective is to enhance the status of aquatic and water dependent terrestrial ecosystems by promoting a responsible water use and reducing pollution. The links between socioeconomic developments and water management are addressed only partially/indirectly.	The Mongolian IWRM policy defines water as state property and grants its citizens "the right to enjoy a healthy and safe environment that is protected against pollution and ecological imbalance". The National Water Resource Plan gathered data about environmental and socio- economic preconditions that is related to water. Nevertheless, a clear implementation strategy is not yet available.
Potential synergies:	
An approach combining the EU-WFD and Mongolian IWRM principles could offer a framework for an environmentally-oriented management of water resources that also incorporates socioeconomic aspects. However, without supplementary implementation strategies emphasizing the social component, the actual focus of the national IWRM concept cannot be implemented sufficiently by only applying European watermanagement strategies.	

Table 4: The First Dublin Principle in the EU-WFD and Mongolian IWRM policy

Water management should be based on a participatory approach, involving users, planners and policymakers at all levels.	
EU-WFD:	**Mongolian IWRM policy:**
The EU-WFD explicitly values public information and consultation but does not explicitly call for public participation in the actual implementation process. RBM drafts and background documents have to be made accessible to the public for at least six months before they are finally adopted. While decentralization and multi-level cooperation are not prescribed, they are indirectly promoted (decision-making processes should be made as locally/regionally as possible).	The Mongolian IWRM policy addresses public participation in the sense of the second Dublin Principle by the introduction of River Basin Councils (RBCs)/ River Basin Administrations (RBAs). RBCs/RBAs are designed to act as a platform to encourage multi-stakeholder discussion as well as information and active involvement of public water users. However, in practice public involvement is often still limited.
Potential synergies:	
Improved information and involvement of the general public would increase transparency and sense of ownership, and thus facilitate a wider public support for IWRM.	

Table 5: The Second Dublin Principle

Women play a central part in the provision, management and safeguarding of water.	
EU-WFD:	**Mongolian IWRM policy:**
The EU-WFD does not consider gender issues. Concerns about gender equality are addressed by the EU's supplementary gender legislation.	The Mongolian IWRM policy does not consider gender issues.
Potential synergies:	
Here, Mongolia cannot benefit from experiences in the European water sector. However, gender equality in Mongolia is, when compared to other transition and developing countries, high in the water sector (Asian Development Bank and World Bank 2005; Hawkins 2007). Therefore, this is not a priority.	

Table 6: The Third Dublin Principle

Water has an economic value in all its competing uses and should be recognized as an economic good.	
EU-WFD:	**Mongolian IWRM policy:**
The EU-WFD regards the enhancement of the ecological status as an investment in future economic benefit. Of special importance are economic analyses to support the introduction of economic instruments that set incentives for a responsible water use, like adjusted cost recovery systems.	The economic aspects of water are addressed by the 'Water Use Right' and the 'Law on Fees for Use of Water Resources and Mineral Water'. However, formulations are often vague and clear implementation strategies are lacking. While there is only a symbolic or even no fee for domestic water use (and thus little incentive to save water), there are plans raise water fees in the future.
Potential synergies:	
The strict guideline for economic analyses provided by the EU-WFD may be helpful for Mongolia. However, a lack of data and non-transparency in parts of the Mongolian water sector are still obstacles.	

Table 7: The Fourth Dublin Principle

3. Conclusion

Even though the EU-WFD cannot be considered a blueprint for IWRM implementation in Mongolia, it can provide guidance in cases where relevant water management policies are still lacking in Mongolia and there is positive experience in Europe.

In general, the EU-WFD serves as a model to enhance the systematic implementation of an environmentally sustainable water resource management at river basin-scale. At first sight, it may seem like the relatively good ecological status of Mongolia's surface water bodies constitutes a much easier starting point when compared to the heavily altered rivers and lakes in Europe that are impacted by multi-stressors. In this light, the EU-WFD objective of reaching and maintaining a good ecological status of all surface water bodies appears to be achievable. The adoption of the EU-WFD would thus be a chance to create a legal framework to maintain the natural water resources and implement a sustainable water use that supports the national IWRM agenda.

However, despite democratization and substantial economic growth over the past decade, Mongolia cannot yet be considered a developed country as it still faces urgent problems related to health, poverty, infrastructure, low employment rates (Porsche-Ludwig et al. 2013) and especially corruption (Fritz 2007; Quah 2006; Transparency International 2012). Moreover, the current pace of the rural exodus and precarious social conditions in the fast growing urban peripheries have no analogy in Europe. Not surprisingly, such challenges were not taken into account when the EU-WFD was drafted.

The EU-WFD addresses only the ecological status of water bodies and although it enforces the identification of pressure sources, it has no direct regulations for urban and economic water use. Without comprehensive supplementary frameworks concerning the restriction of point and diffuse pollutions at its sources, significant improvements are not to be expected. Socioeconomically, Mongolia has the closest similarities with the formerly socialist Eastern European countries and shares the challenge of water scarcity with Southern European nations. Not all EU countries, as for example Portugal and Greece, have been able to submit their RBMPs to the EU within the prescribed deadline (EC 2012e; EC 2012f) even though they received considerable support from the EU. In contrast to these countries, Mongolia as a non-EU Member State is not able to receive similar long-term funding and support in implementing the extensive requirements such as establishing a sufficient monitoring network or eliminating deficits in the domestic water sector. Nevertheless, the mentioned environmental and socio-economic similarities and comparable anthropogenic impacts in some regions of the EU and Mongolia mean that strategies, methodologies and metrics from Europe may be partially transferable.

Some differences do however significantly limit the usability of the EU-WFD as a guideline for IWRM implementation in Mongolia. One important difference is related to challenges posed by the legal framework and financial and institutional capacities, for which the EU-WFD does not provide an easy solution. Moreover, European water management strategies are the result of a highly dynamic scientific sector. New developments that focus on the incorporation of ecosystem services, ecosystem functioning and environmental risk assessment should not be neglected, as they may offer an adequate alternative that is even more suitable to enhance the implementation of context specific sustainable water resources management in non-European countries and thus promote the implementation of a comprehensive IWRM.

References

Asian Development Bank, World Bank (2005) Mongolia Country Gender Assessment. Manila, Philippines.

European Commission (2008) EU Gender Equality Law. Luxembourg.

European Commission (2012a) Member State Finland: report from the Commissione toe the European Parliament and the Council on the implementation of the Water Framework Directive (2000/60/EC) – River Basin Management Plans. Brussels, Belgium.

European Commission (2012b) Member State Malta: report from the Commissione toe the European Parliament and the Council on the implementation of the Water Framework Directive (2000/60/EC) – River Basin Management Plans. Brussels, Belgium.

European Commission (2012c) Member State the Netherlands: report from the Commissione toe the European Parliament and the Council on the implementation of the Water Framework Directive (2000/60/EC) – River Basin Management Plans. Brussels, Belgium.

European Commission (2012d) Member State Romania: report from the Commissione toe the European Parliament and the Council on the implementation of the Water Framework Directive (2000/60/EC) – River Basin Management Plans. Brussels, Belgium.

European Commission (2012e) Member State Porutgal: report from the Commissione toe the European Parliament and the Council on the implementation of the Water Framework Directive (2000/60/EC) – River Basin Management Plans. Brussels, Belgium.

European Commission (2012f) Member State Greece: report from the Commissione toe the European Parliament and the Council on the implementation of the Water Framework Directive (2000/60/EC) – River Basin Management Plans. Brussels, Belgium.

European Environmental Bureau (Hrsg.) (2010) 10 years of the Water Framework Directive: A Toothless Tiger?. Brussels, Belgium.EU-Water Framework Directive (EU-WFD) 2000. Richtlinie 2000/60/EG des Europäischen Parlamentes und Rates vom 23. Oktober 2000 zur Schaffung eines Ordnungsrahmens für Maßnahmen der Gemeinschaft im Bereich der Wasserpolitik.

Flussgebietsgemeinschaft Elbe (2009) Verwaltungsvereinbarung über die Gründung einer Flussgebietsgemeinschaft für den deutschen Teils des Einzugsgebietes der Elbe (FGG Elbe). Wedel, Germany.

Fritz V (2007) Democratization and Corruption in Mongolia. Public Administration and Development 27(3): 191–203.

Genté R (2013) Die Jurtenviertel von Ulan-Bator. LE MONDE diplomatique 08/03/2013.   http://www.mondediplomatique.de/pm/2013/03/08.mondeText.artikel,a0048.idx,14

Global Water Partnership (2009) A Handbook for Integrated Water Resource Management in Basins. www.gwpforum.org.

Grambow M (Hrsg.) (2013)Nachhaltige Wasserbewirtschaftung. Springer Vieweg, Wiesbaden.

Hartwig M, Theuring P, Rode M, Borchardt D (2012) Suspended sediments in the Kharaa River catchment (Mongolia) and its impact on hyporheic zone functions. Environ Earth Sci 65(5):1535-1546. doi: 10.1007/s12665-011-1198-2

Hawkins R (2007) Gender and Water Resource Management in Mongolia. Master Thesis, University of York Toronto, Canada.

Heldt S (2014) The EU-WFD as an Implementation Tool for IWRM in non-European Countries – Case Study: Drafting a River Basin Management Plan for the Kharaa River Basin in Nothern Mongolia. Master Thesis. University of Duisburg-Essen, Essen, Germany.

Hering D , Borja A, Carstensen J, Carvalho L, Elliott M, Feld CK, Heiskanen AS, Johnson RK,Moe J, Pont D, Solheim AL, van de Bund W (2010) The European Water Framework Directive at the age of 10: A critical review of the achievements with recommendation fort he future. Science of the Total Environment 408(19): 4007-19.

Hofmann J, Hürdler J, Ibisch R, Schaeffer M, Borchardt D (2011) Analysis of Recent Nutrient Emission Pathways, Resulting Surface Water Quality and Ecological Impacts under Extreme Continental Climate: The Kharaa River Basin (Mongolia). Internat Rev Hydrobiol 96(5): 484-519. doi: 10.1002/iroh.201111294

Hofmann J, Venohr M, Behrendt H, Opitz D (2010) Integrated water resources management in central Asia: nutrient and heavy metal emissions and their relevance for the Kharaa River Basin, Mongolia. Water Sci Technol 62(2):353-363. doi: 10.2166/wst.2010.262.

Hofmann J, Watson V, Scharaw B (2014): Groundwater quality under stress: contaminants in the Kharaa River basin (Mongolia). Environ Earth Sci, this issue. doi: 10.1007/s12665-014-3148-2

Houdret A, Dombrowsky I, Horlemann L (2013) The institutionalization of river basin management as politics of scale - insights from Mongolia. Journal of Hydrology, in Press, Corrected Proof.

International Office for Water (2009) Capacity building for a better water management – Organization of water management in France. Paris, France.

Janzen J (2012) Die Mongolei im Zeichen von Marktwirtschaft und Globalisierung. Geographische Rundschau 64(12):4-10.

Karthe D, Heldt S, Houdret A, Borchardt D (2014b): IWRM in a country under rapid transition: lessons learnt from the Kharaa River Basin, Mongolia. Environmental Earth Sciences, submitted.

Karthe D, Heldt S., Rost G, Ilian J, Heppeler J, Sullivan C, Stäudel J, Scharaw B, Westerhoff T, Sigel K, Watson V, Borchardt D (2014b): Modular Concept for Municipal Waste Water Management in the Kharaa River Basin, Mongolia. Submitted to Borchardt D, Ibisch R (2014): Integrated Water Resources Management: Concept, Research and Implementation.

Karthe, D.; Kasimov, N.; Chalov, S.; Shinkareva, G.; Malsy, M.; Menzel, L.; Theuring, P.; Hartwig, M.; Schweitzer, C.; Hofmann, J.; Priess, J. & Lychagin, M. (2014c): Integrating Multi-Scale Data for the Assessment of Water Availability and Quality in the Kharaa - Orkhon - Selenga River System. Geography, Environment, Sustainability 3(7):65-86.

Karthe D, Sigel K, Scharaw B, Stäudel J, Hufert F, Borchardt D (2012): Towards an integrated concept for monitoring and improvements in water supply, sanitation and hygiene (WASH) in urban Mongolia. Water & Risk 20:1-5.

Minderlein S, Menzel L (2014): Evapotranspiration and energy balance dynamics of a semi-arid mountainous steppe and shrubland site in northern Mongolia. Environ Earth Sci. doi: 10.1007/s12665-014-3335-1

Mongolian Ministry for Environment and Green Development (MEGD) (2012) Integrated Water Resource Management National Assessment Report - Vol. I & Vol. II. Ulaanbaatar, Mongolia.

ORSAM (2011) EU's Water Framework Directive Implementation in Turkey: The Draft National Implementation Plan. Ankara, Turkey.

Pfeiffer M, Batbayar G, Hofmann J, Siegfried K, Karthe D, Hahn-Tomer S (2014): Investigating arsenic (As) occurrence and sources in ground, surface, waste and drinking water in northern Mongolia. Environ Earth Sci. doi: 10.1007/s12665-013-3029-0

Porsche-Ludwig, M., Gieler, W., Bellers, J. (2013) Handbuch Sozialpolitik der Welt, LIT-Verlag, Berlin.

Quah JST (2006) Curbin Asian Corruption: An Impossible Dream?. Current History 105: 176-179.

Rahaman MM, Varis O, Kajander T (2004) EU Water Framework Directive vs. International Principles Concerning IWRM: the Seven Mismatches. International Journal of Water Resource Development 20(4): 565- 575.

Schweitzer C (2012) Modelling land-use and land-cover change and related environmental impact in Northern Mongolia. Dissertation. Martin-Luther-University Halle-Wittenberg, Halle (Saale), Germany.

Sigel K, Altantuul K & Basandorj D (2012) Household needs and demand for improved water supply and sanitation in peri-urban ger areas: The case of Darkhan, Mongolia. Environ Earth Sci 65(5):1561-1566. doi: 10.1007/s12665-011-1221-7

Snellen WB, Schrevel A (2004) IWRM: for sustainable use of water – 50 years of international experiences with the concept of integrated water management, Netherlands Conference on Water for Food and Ecosystems, 2004, Wageningen.

Stalnacke P, Gooch GD (2010) Integrated Water Resource Management. Irrig Drainage Syst (2010) 24:155–159.

Sumer V, Muluk C (2011) Challenges for Turkey to implement the EU Water Framework Directive. Turkey's Water Policy: 43-67. Heidelberg, Germany.

Theuring P, Rode M, Behrens S, Kirchner G, Jha A (2013) Identification of fluvial sediment sources in a meso-scale catchment, Northern Mongolia. Hydrol Process 27(6):845-856. doi: 10.1002/hyp.9684

Thorslund J, Jarsjö J, Belozerorva E, Chalov S (2012) Assessment of the gold mining impact on riverine heavy metal transport in a sparsely monitored region: the upper Lake Baikal Basin case. J Environ Monit 14(10):2780-2792. doi: 10.1039/C2EM30643C

Uddin SNM, Li Z, Gaillard JC, Tedoff PF, Mang HP, Lapegue J, Huba EM, Kummel O, Rheinstein E (2014): Exposure to WASH-borne hazards: A scoping study on peri-urban Ger areas in Ulaanbaatar, Mongolia. Habitat Int 44:403-411.

United Nation Development Programm (UNDP) & United Nation Children's Fund (UNICEF) (2009) Improving local services for the Millennium Development Goals – Rural Water Suppla and Sanitation in Mongolia. Ulaanbaatar, Mongolia.

United Nation Development Programm (UNDP) (2014) Human Development Report 2014 - Sustaining Human Progress: Reducing Vulnerabilities and Building Resilience. New York, USA.

Transparency International (2012) Corruption in natural resource management in Mongolia. www.U4.no

Women in Europe for a Common Future (2008) Europe's Sanitation Problems. Report of the World Water Week Seminar. Stockholm, Sweden.

World Health Organization (2010) Health and Environment in Europe: Progress Assessment. Copenhagen, Denmark.

Potential and feasibility of willow vegetation filters in Mongolia

Katja Westphal[a], Chris Sullivan[a], Peder Gregersen[b], Daniel Karthe[a]

[a] Helmholtz Centre for Environmental Research, Leipzig/Magdeburg,
Germany. Corresponding author: katja.westphal@posteo.de
[b] Center for Recirkulering, Ølgod, Denmark

Abstract

In Mongolia, the discharge of insufficiently treated wastewater to the receiving environment due to poor connection rates of informal ger areas to wastewater facilities is a significant problem. To address nutrient emission problems as well as hygiene issues, we investigated the potential of willow vegetation filters in the Kharaa River Basin in Mongolia by (1) testing a pilot willow wastewater treatment plant and (2) developing different design and cost scenarios of a large willow vegetation filter for the model village of Khongor in order to identify a suitable filter type. This work was carried out in the context of a research and development project aiming at the development and implementation of an IWRM concept for the Kharaa River Basin.

Initial results of the pilot plant show that willow vegetation filters are an appropriate way to treat wastewater in Mongolia since treatment performance and operation of the system are satisfying, even under very cold conditions. Moreover, the risk for contaminating the groundwater and soil matrix was found to be low.

Among three alternatives, open evapotranspirative vegetation filters were identified as the most appropriate solution. These systems are not lined and allow some percolation of irrigation water into deeper soil zones. Decisive criteria are next to very low life cycle costs, minimum health risk by subsurface irrigation, high operation reliability and comparatively low land requirements.

1 Introduction

Mongolia is faced with far-reaching changes like land use intensification and increasing urbanization. In addition, larger numbers of livestock, the expansion of mining industries and climatic changes such as increasing temperatures, higher evapotranspiration rates and expected reductions in precipitation are putting pressure on the nation's water resources. These developments show the necessity to develop effective methods in order to protect the region's water quality and quantity under the changing social, demographic and environmental conditions (Batimaa et al. 2008). For these reasons, Mongolian and German research institutes have launched the IWRM MoMo project, dedicated to develop and implement strategies leading towards an integrated water resource management for the Kharaa River Basin - a Mongolian model region (Karthe et al. 2014).

One fundamental problem identified during the two project phases is the discharge of insufficiently treated wastewater to the environment. It is assumed that only 35 % of the total population in the river basin is connected to wastewater facilities that are mostly in a poor state of repair (Hofmann et al. 2011). The discharge of untreated wastewater mainly from informal ger areas in urban settlements represents a considerable part of the total nutrient emissions within the basin. Negative impacts on public health associated with untreated human waste are presumed, including a high prevalence of gastrointestinal infections including Hepatitis A (Sigel et al. 2011; Karthe et al. 2012).

To address nutrient emission problems as well as possible hygiene issues, the establishment of reliable wastewater treatment facilities is an important requirement of an effective IWRM approach in the Kharaa River Basin. A major challenge for the reliable and cost-effective operation of wastewater infrastructures is the harsh Mongolian climate. Very cold temperatures of less than -30° C can have significant impact on hydraulic and biological performances of wastewater treatment systems. Hence, conventional methods for treating wastewater may not be appropriate.

In order to address these challenges of wastewater treatment in Mongolia, the project team investigated the possibility to establish a willow vegetation filter in the Kharaa River Basin. These decentralized facilities are comparable to slow rate land treatment systems, where wastewater is applied on land planted with willows or other fast growing crops such as poplars. By applying wastewater on short rotation coppice, sewage can be effectively treated by physical, chemical and biological processes within the soil-plant-water matrix while also considerably promoting biomass production, even in cold climate. Thereby, wastewater becomes an important resource rather being a subject of concern. In addition, willow vegetation filters provide the advantages of being inexpensive, simple and robust. In this way the system addresses Mongolian problems of insufficient sanitation, growing water scarcity and progressive deforestation resulting in shortage of timber. For these reasons, the project team investigated the potential and feasibility of willow vegetation filters for Mongolia by:

- Testing a pilot willow wastewater treatment plant at the Technical University in Darkhan (MUST)
- Developing different design and cost scenarios of a large-scale willow vegetation filter for the model village of Khongor

2 Pilot willow wastewater treatment plant

To investigate the feasibility of willow vegetation filters in Mongolia, a willow wastewater treatment facility was developed and installed on the Campus of the Mongolian University of Science and Technology (MUST), modeled on a Zero Discharge System according to Gregersen and Brix (2001). Pre-treated wastewater irrigated four beds of 16 m^2 that were lined with a PVC membrane and planted with one year old endemic poplar and *salix* species. Drainage water running from the beds by gravity was then analysed for the following physico-chemical parameters on a weekly basis: chemical oxygen demand (COD), biological oxygen demand (BOD_5), total phosphorous (TP), total nitrogen (TN), ammonium, nitrate, nitrite and E. coli. The volume of leachate was measured by tipping buckets (Karthe et al. 2015).

In order to investigate the influence of loading rates and irrigation cycle on water quality and balance as well as on tree vitality and biomass production, different irrigation periods and wastewater loads were tested. For a schematic diagram and further details on the experimental design, see Fig. 1.

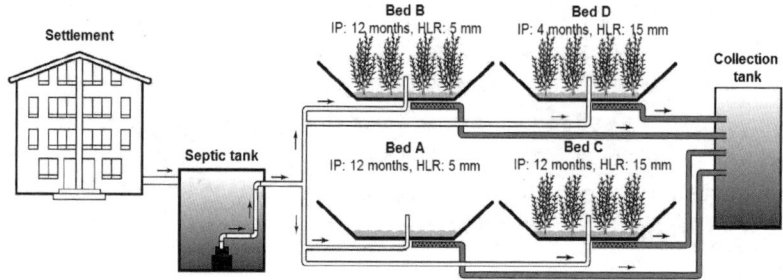

Fig. 1: Schematic diagram of the trial system and the experimental design, including the temporal irrigation pattern and irrigation load. (abbr.: IP - irrigation period, HLR - hydraulic loading rate) (Karthe et al. 2015).

Initial results of treatment efficiency of the system had proved to be satisfactory for the first year of operation in 2011: BOD_5 - removal rates ranges between 79 % and 93 %, while rates for COD vary between 60 % and 87 %. The highest removal rates for TN and TP were achieved by bed B with 77 % and 89 %, respectively. Removal rates for TN were lowest for bed C (38 %). Lowest removal rates for TP amount to only 42 % in bed D (see Tab. 2). Results of the second year of operation are even more promising. The estimated biomass production for the first year of growing ranges between 11 t to 24 t DM ha^{-1} yr^{-1} (Karthe at al. 2015).

Bed	BOD_5 [%]	COD [%]	TN [%]	TP [%]
Bed A	79	60	53	61
Bed B	93	87	77	89
Bed C	87	70	38	43
Bed D	89	67	63	42

Tab. 2: Treatment performances of the four trial beds

The storage of wastewater within the willow and poplar beds during the winter months has proven to be successful, since trees were still vital in spring. Additional housing or wastewater storage is therefore not necessary. However, in spring higher quantities of percolation water are generated due to snowmelt (Karthe et al. 2015)

3 Development of different designs and cost scenarios

The results of testing the pilot plant clearly showed that willow vegetation filters are an appropriate way to treat wastewater in Mongolia since purification capacity and operation of the system are satisfying, even in very cold months. Moreover, the risk for contaminating groundwater and soil matrix is low. However, the question of a suitable design and type of vegetation filter remains open as there is a variety of systems that are already being used in different parts of the world. Since lined (i.e. sealed) systems such as tested in the pilot plant in Khongor are on a large scale basis rather expensive, 'open systems' which have been tested in several countries including Sweden and China are an alternative (Wu et al. 2011, Ou et al. 1997, Qu-Xing et al. 2006, BIOPROS 2008). In order to identify a suitable willow filter type for Mongolia, three scenarios for the model village of Khongor were developed considering possible designs, sizes and costs. Based on various criteria, a final proposal was made for an appropriate type of system by means of a utility value analysis (see also Westphal 2013). The evaluation and identification was carried out for three willow filter systems that were identified in the literature as the most common types but also occur in modified versions in practice.

The first type is a surface irrigated short rotation willow coppice (SRC) whose wastewater is stored during winter in a pond. Irrigation pauses during cold months. SRCs are not lined. Hence, some percolation in deeper soil layers and groundwater is possible. The second type is an evapotranspirative system with year-round subsoil irrigation. Storage of the water during colder months with low evapotranspiration occurs in the soil matrix within the willow beds. Evapotranspirative systems are constructed in two versions. Zero discharge systems (ZDS) are enclosed either with a water impermeable layer. Open systems (OS) do not have any liner. They allow some percolation. Usually, open systems are used for areas of low groundwater interests. However, both types fulfill highest Danish standards of wastewater management (Gregersen 2013).

3.1 Methods

3.1.1 Site description and assumptions

Khongor is approximately 200 km north of Ulanbataar and 20 km south of Darkhan. The region is characterized by very cold winters and warm short summers. Mean annual air temperatures are around 0°C. Average annual precipitation accounts for only 270 mm (WMO 2013) and falls far below annual potential evapotranspiration of around 800 mm (Karthe et al. 2011). Actual annual water requirements by willows (ET_C) were calculated with CROPWAT (by FAO) and amount to 1330 mm (see Guidi et al. 2007). The groundwater table in the area lies 4 m to 5 m below the surface. The unsaturated soil zone consists of loamy-sandy sediments (Hofmann 2008). Kastanozem was identified as the dominant soil type (MoMo 2009). In order to facilitate the designing and evaluation process, assumptions were made (see Tab. 3).

	Design criteria	Value	Reference
General	Population equivalent	3000 PE	Authors' assumption
	Water consumption	170 l d^{-1}	Authors' assumption
	N_{tot} load p.p. after PT	10 g d^{-1}	Gujer 1999
	P_{tot} load p.p. after PT	2,5 g d^{-1}	Gujer 1999
	Wood moisture rate	15 %	Grosser 2005
	Useable fract. of wood	80 %	Authors' assumption
	Gross density of wood	560 kg/m^3	Grosser 2005
SRC	NLR	250 kg ha^{-1}yr^{-1}	Aronsson & Bergström 2001
	PLR	60 kg ha yr^{-1}	Aronsson & Bergström 2001
	Biomass production	10 t DM ha^{-1} yr^{-1}	Aronsson & Perttu 2001
ZDS	WSC ≙ pore volume	40 %	Authors' assumption
	Biomass production	15 t DM ha^{-1} yr^{-1}	Authors' assumption
OS	WSC ≙ field capacity	30 %	Authors' assumption
	Leaching requirement	15 %	Asano et al. 2006

Irrigation efficiency	80 %	Asano et al. 2006
DVF	20 %	Asano et al. 2006
N-uptake by crop	170 kg ha^{-1} yr^{-1}	Authors' assumption
N-conc. in percol. water	15 mg N$_{tot}$ l^{-1}	Hofmann, pers. com. 2013
Biomass production	15 t DM ha^{-1} yr^{-1}	Authors' assumption

Tab. 3: Assumption list as a basis of computation for design and costs (abbr.: PE - personal equivalent, p.p. - per person, PT - pretreatment   NLR - nitrogen loading rate, PLR - phosphorous loading rate, DM - dry matter, WSC - water storage capacity), DVF - Denitrification and volatisation fraction)

3.1.2 Size determination

For determining the size of the SRC system, the recommendations by Aronsson and Bergström (2001) were followed. Based on their work, the maximum allowable loading rate for nitrogen is defined as 250 kg N ha^{-1} yr^{-1}. Maximum allowable loading rate for phosphorous is defined as 60 kg P ha^{-1} yr^{-1}. From the annual amount of wastewater and its total nitrogen and phosphorous load, the size of the system can be determined. The plantation's design follows for the most part recommendations of BIOPROS (2008).

The process of size determination for ZDS and OS is based on the difference between annual precipitation and ET$_C$ for willows. This difference is equal to the annual amount of sewage that can be applied to the willow beds (see also Brix & Arias 2011, Gregersen & Brix 2001). The thereby calculated area represents the minimum size of the systems. Since seasonal climatic variations are not considered, it must be ensured that the willow beds have enough soil pore volume to store rain and wastewater in winter, when rates for evapotranspiration are not sufficient. 40 % pore volume for the ZDS and 30 % pore volume (corresponding to approximate field capacity for reasons of groundwater interest) for the OS was the basis for calculating the water storage capacity of the soil. Therefore, a balance calculation of the inflow parameters (P, WW input) and the outflow parameters (irrigation load, ET) on a monthly basis was performed to determine the maximum accumulating water volume. To ensure a minimum risk to groundwater, two additional control calculations according to Asano et al. (2006) were performed (L$_{W(1)}$ and L$_{W(N)}$). Details are shown in

Tab. 3. The system's designs mainly follow general recommendations of Peder Gregersen (pers. com. 2013).

3.1.3 Cost Calculation

Cost calculations for each design proposal are an important part of the evaluation and identification process. Therefore, not only the investment costs play an important role, but also operation costs, and profit gained by selling the harvested biomass. All computed costs and profit are based on a price list for material, construction works, operation and maintenance compiled by Mongolian contractors associated with the MoMo project and were depreciated with a rate of 5 % over a lifespan of 25 years (BIOPROS 2008, Börjesson & Berndes 2006).

The costs of investment for all three systems include all bigger material and construction items such as distribution pipework, synthetic liner, willow cuttings, planting costs or earthworks, including replacement items such as fences or pumps. An estimate of operation costs includes works like harvesting, weeding or pump maintenance. For estimating the profit gained by selling the willows as firewood it was necessary to make assumptions in terms of biomass production, wood moisture and gross density (see Tab. 3). Eventually, total lifecycle costs results from the sum of investment, operation and replacement cost, less profit gained by selling the firewood.

3.1.4 Utility value analysis (UVA)

To evaluate the suitability of the three suggested systems, a utility value analysis was undertaken. This form of analysis is an assessment tool from the economic sector to evaluate not only monetary factors but also the benefit of projects among a number of different alternatives (Burghardt 2007). The UVA in this work was carried out on the basis of "Method and Techniques of Organization" by Schmidt (2000). Assessment criteria include considerations about nutrient application rates, design criteria, cost associated criteria and risk of contamination to public health and environment (see Tab. 6).

3.2 Results and Discussion

3.2.1 Dimensioning and design of the systems

The size of the SRC was calculated to be 43.8 ha. The total area of the willow plantation has a size of 790 m x 586 m, including machinery access roads. The SRC-system consists of 8 irrigation blocks. Each irrigation block has 256 rows of willows. Two willow rows will be irrigated by one lateral. For further characteristics of the system see Tab. 4.

The size calculation of the ZDS based on evapotranspiration resulted in a minimum area of 18.5 ha. Since the soil volume underneath cannot absorb the full amount of wastewater the area has to be enlarged to 19.08 ha. The system exists of 6 irrigation blocks, which consists of 22 lined willow beds with an unplanted 5 m-corridor in-between. Every bed has 6 rows of willows. One willow bed is provided with one lateral for irrigation. The total size of the system amounts to 608 m x 602 m. Further details are provided in Tab. 4.

The size of the OS had to be enlarged from originally 18.5 ha to 24.8 ha for reasons of sufficient water storage capacity of the soil. To further ensure that there is a minimum risk to groundwater, control calculations based on $L_{W(1)}$ and $L_{W(N)}$ according to Asano et al. (2006) resulted in sizes of 14.25 ha and 18.57 ha, respectively. For reasons of groundwater protection the biggest area of 24.8 ha was subject of a further design process. The subsequent designed OS consist of 8 irrigation blocks with 22 non-lined willow strips, respectively. Planting and general design principles follow the example of the ZDS. The irrigation is carried out by two underground laterals per willow strip to ensure an even distribution of wastewater. In this way the total size of the system amounts to 779 m x 602 m. For further characteristics and a draft of the system see Tab.

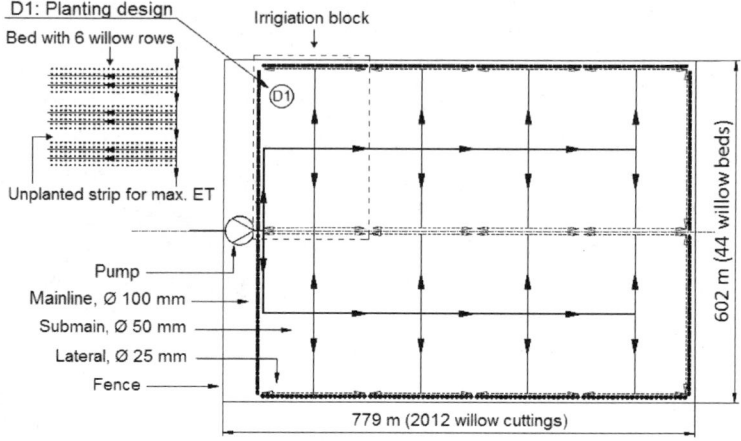

4 and

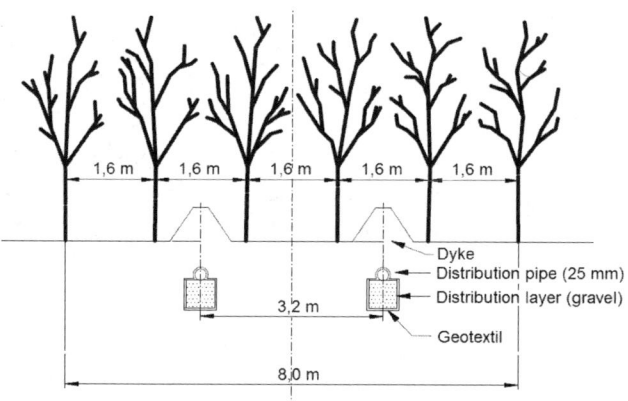

Fig. 2, respectively.

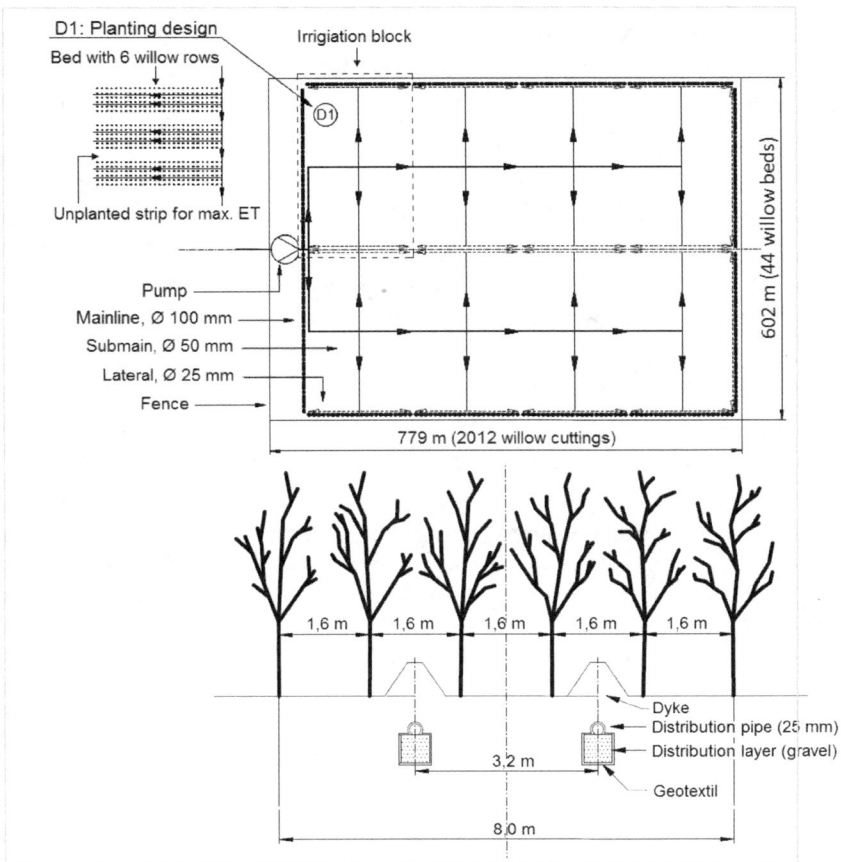

Fig. 2: Schematic diagram of the OS for Khongor (above) and a cross section through a willow bed (below) (abbr.: ET- Evapotranspiration).

Since there are limited experiences with SRC systems in very cold climates like Mongolia, size comparisons for a plausibility check to other already established systems are very difficult. However, examples from Inner Mongolia and Sweden show that comparable treatment systems require a similar irrigation area per treated m^3 of wastewater (Ou et al. 1997, BIOPROS 2008). While a system for Khongor would need an amount 2.35 m^2, a wastewater irrigated forest from China needs 2.41 m^2 per treated m^3 of wastewater (Ou et al. 1997).The plausibility check for the ZDS and OS sizes is difficult since these designs depend on local climate. In addition experiences described in the literature are very rare. However, Brix & Arias (2011) and Gregersen & Brix (2001) describe sizes of 0.7 m^2 to 1.6 m^2 per treated m^3 of wastewater in Denmark. The calculated sizes for the ZDS and OS range between 1.0 m^2 and 1.3 m^2 per treated m^3 of wastewater, respectively. Hence, sizes for ZDS and OS are in a middle range, compared to Danish values.

The salinisation risk of the onsite soil remained unconsidered. Kastanozems in particular are sensitive to secondary salinisation when irrigated (ISRIC 2013). Therefore, it is necessary to compute leaching requirements for salinity control by means of Asano et al. (2006) and USDA (1993). Ou et al. (1997) and Qu-Xing et al. (2006) also discussed salinisation of wastewater irrigated land. Due to a lack of information, drainage and leaching requirements for salinity control was not computed for Khongor but is strongly recommended for future works. Because willows are very tolerant to high salt concentrations it is not expected that salt accumulation in the soil will have adverse effects on biomass yield and health of willows (Hangs et al. 2011, Brix & Arias 2011, Gregersen pers. com. 2013).

General Characteristics	SRC	ZDS	OS
Irrigation area [ha]	43.80	19.08	24.80
Total area [ha]	61.53	35.70	45.82
Winter storage [ha]	3.86	within soil	within soil
Irrigation days [d]	153	365	365
HLR [mm d^{-1}]	2.50	2.67	2.06
NLR [kg N ha^{-1} yr $^{-1}$]	250	574	442
PLR [kg P ha^{-1} yr $^{-1}$]	58	132	102
Area per person [m^2 person^{-1}]	205	119	153
Distribution. Pipework-25 mm [m]	194,560	23,892	61,952
Distribution pipework-50 mm [m]	2,304	1,686	2,248
Distribution pipework-100 mm [m]	1,708	1,251	1,598
Earthworks [m^3]	105,110	233,994	39,094
Liner synthetic [m^2]	40,869	238,487	-
Geotextil [m^2]	81,738	529,536	136,294
Gravel 20mm [m^3]	-	5,973	15,488
Pump housing [pce]	1	1	1
Number of pump [pce]	1	1	1
Fencing [m]	3,332	2,420	2,762
Willlow cuttings [pce]	647,168	409,464	531,168
Plant density [pce ha^{-1}]	14,776	21,460	21,420

Tab. 4: General design characteristics of the 3 systems (abbr.: HLR- hydraulic loading rate, NLR - nitrogen loading rate, PLR – phosphorous loading rate)

3.2.2 Cost Calculation

The initial investment costs for the SRC-system are estimated at 1,840,697 €. Most of the costs are related to earthworks for the wastewater storage pond

and the synthetic liner. The total initial investment costs for the ZDS are estimated at 4,065,114 €. The biggest share of initial costs is also attributed to the
synthetic liner and earthworks. The initial investment costs for the OS amount
to 810,760 €. Most of the costs are associated to the willow cuttings, earthworks and gravel, which is needed as a distribution layer for the irrigation water.
For an overview of the total investment costs see Fig. 3 .

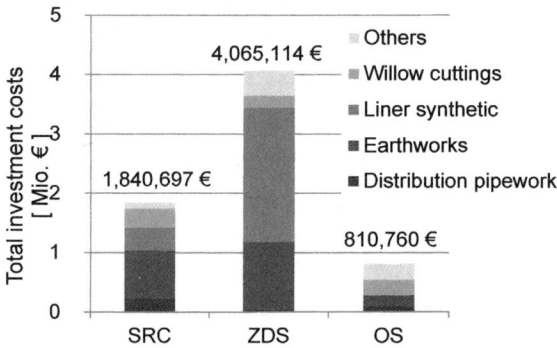

Fig. 3: Expected initial investment costs for all three systems. The costs for investment are five
times higher than for the OS. Costs for the SRC system are over two times more expansive than
the OS.

The expected life cycle costs only slightly exceed the cost for investment, since
replacement and operation costs are expected to be relatively low. In this way
the profit, made by selling the firewood, almost balance the costs of operation.
For a more detailed cost listing see Tab. 5.

However, the determined life cycle costs for all three systems can only
be a rough estimation since the cost computation is based on very simplified
designs and a set of assumptions (see Tab. 3). Another important cost-related
aspect that remained unconsidered, but can be crucial, is the necessity for primary treatment. While SRC systems include a wastewater storage pond that
works as an anaerobic settling pond, additional primary treatment systems are
required for a ZDS and OS.

Costs	SRC	ZDS	OS
Initial investment (-)	1,840,697 €	4,065,114 €	810,760 €
Replacement (-)	102,596 €	91,388 €	95,591 €
Operation (-)	508,850 €	294,736 €	332,343 €
Profit (+)	410,113 €	228,183 €	348,046 €
Total	2,042,029 €	4,223,055 €	890,648 €

Tab. 5: Life cycle cost for the systems, including profit made by selling the harvested wood.

3.2.3 Utility value analysis

The utility value analysis identified the OS as the most suitable system for Khongor with a value of 217 points. Key advantages are its relatively low land requirements and its very low investment costs. Hardly any earthworks are required and no synthetic membrane is needed. Next to cost advantages, the system is expected to be very reliable and enables a minimum health risk due to subsurface irrigation. Tab. 6 gives an overview of the applied criteria, their weighting and the eventual distribution of points.

However, the analysis contains a significant degree of subjectivity due to point awarding, weighting and selection of criteria. A change in weighting and point awarding or the additional evaluation of criteria can lead to a different result, especially taking into account that results are very close to each other.

Criteria	Weight. [%]	SRC	ZDS	OS
Nutrient application - NLR, PLR	20	60	20	20
Design criteria - e.g. land requirements	20	33	51	51
Cost criteria - investment, operation, profit	30	59	42	86
Risk criteria - human health, groundwater	30	45	90	60
Total	100	197	203	**217**

Tab. 6: Overview of the utility value analysis. The highest weighting received cost and risk criteria (30 %), followed by criteria for nutrient application and design (20 %). The results of the analysis are very close together.

4 Conclusions

In order to address the challenge of insufficiently treated wastewater in Mongolia, the project team investigated the potential and feasibility of willow vegetation filters by testing a pilot willow wastewater treatment plant and developing different design and cost scenarios for a large scale vegetation filter. Due to the satisfying purification capacity and operation of the pilot plant, willow vegetation filters are considered to be suitable systems for very cold regions like Mongolia. However, lined systems modeled on the pilot plant's design are expensive on a large scale basis. In order to identify an alternative, three scenarios for the model village of Khongor were developed considering possible designs, sizes and costs.

Based on various criteria open evapotranspirative vegetation filters were identified as the most appropriate willow vegetation filter type. These systems are not lined and allow some percolation of irrigation water into deeper soil zones. Crucial criteria are very low life cycle costs, high operation reliability, low land requirements and minimum health risk due to subsurface irrigation.

Acknowledgements

The results presented in this paper are based on the research and development project „Integrated Water Resources Management in Central Asia: Model Region Mongolia", funded by the German Federal Ministry of Education and Research (BMBF) in the framework of the FONA (Research for Sustainable Development) initiative (grant no. 033L003). We acknowledge the support provided by the Project Administration Jülich (PTJ) and the BMBF/International Bureau in the context of the "Assistance for Implementation" (AIM) scheme.

References

Aronsson, P., Bergström, L. F. (2001). Nitrate leaching from lysimeter-grown short-rotation willow coppice in relation to N-application, irrigation and soil type. In: Biomass and Bioenergy, pp. 155–164.

Aronsson, P., Perttu, K. L. (2001). Willow vegetation filters for wastewater treatment and soil redmediation combined with biomass production. In: The Forestry chronicle, Canadian Institute of Forestry, pp. 293-300.

Asano, T., Burton, F., Leverenz, H., Tsuchihashi, R., Tchbanoglous, G. (2006). Water reuse, Issues, Technologies, and Application. New York: McGraw-Hill.

Batimaa, P., Batnasan, N., Bolormaa, B. (2008). Climate change and water resources in Mongolia. In: Basandorj, B. & Oyunbaatar, D. (Ed.) (2008): International Conference "Uncertainties in water resource management: causes, technologies and consequences", pp. 7-12. Jakarta: IHP Technical Documents in Hydrology No. 1.

BIOPROS. (2008). Short Rotation Plantations – Guidelines for the safe application of wastewater and sewage sludge, published by: Ulsters Farmers Union.

Börjesson, P., Berndes, G. (2006). The prospect for willow plantations for wastewater treatment in Sweden. In: Biomass and Bioenergy, pp. 428–438.

Brix, H., Arias, C. A. (2011). Use of Willow in Evapotranspirative Systems for Onsite Wastewater Management – Theory and Experiences from Denmark. Novi Sad: STEPOW.

Burghardt, M. (2007). Einführung in Projektmanagement - Definition, Planung, Kontrolle, Abschluss. Erlangen: Publics Corporate Publishing.

Gregersen, P. (2013). Willow facilities, Center for Recirkulering. Power Point Presentation.

Gregersen, P., Brix, H. (2001). Zero-discharge of nutrients and water in a willow dominated constructed wetland. In: Water Science and Technology, 44 (11–12), pp. 407–412.

Grosser, D. (2005). Das Holz der Weide- seine Eigenschaften und Verwendung, Beiträge zur Silberweide, LWF-Bericht Nr. 24 - Kapitel 15, online available: http://www.lwf.bayern.de/mam/cms04/forsttechnik-holz/da-teien/w24das_holz   _der_weide-seine_eigenschaften_und_verwen-dung.pdf, date: 14.12.2014.

Guidi, W., Piccioni, E., Bonari, E. (2008). Evapotranspiration and crop coeffi-cient of poplar and willow short-rotation coppice used as vegetation filter. In: Bioresource Technology, 99, pp.4832–4840.

Gujer, W. (1999). Siedlungswasserwirtschaft. Springer.

Hangs, R. D., Schoenau, J. J., Van Rees, K. C. J., Steppuhn, H. (2011). Ex-amining the salt tolerance of willow (*Salix* spp.) bioenergy species for use on salt-affected agricultural lands. In: Canadian Journal of Plant Science, 91(3), pp. 509-517.

Hofmann, J. (2008): Bericht über die Untersuchungen von Grundwasser und Boden auf Schwermetalle und Cyanid in Khongor Sum, Verbundfor-schungsvorhaben „Integriertes Wasserressourcen- Management in Zentralasien: Modellregion Mongolei (MoMo)", Berlin: IGB.

Hoffmann, H., Platzer, C., Winkler, M., von Muench, E. (2011). Technology re-view of constructed wetlands, Subsurface flow constructed wetlands for greywater wand domestic wastewater treatment, published by: Deutsche Gesellschaft für Internationale Zusammenarbeit (GIZ) GmbH.

ISRIC (2013): International Soil Reference and Information Centre, Kastano-zems (KS), onine available: http://www.isric.org/isric/webdocs/docs//ma-jor_soils _of _the_world/set8/ks/ kastanoz.pdf, date: 12.12.2014.

Karthe, D., Borchardt, D., Kaus, A. (2011). Towards an Integrated Water Re-sources Management for the Kharaa Catchment, Mongolia. In: Гуринович, А.Д. (Hrsg.) (2011): Proceedings of the IWA 1st Central Asian Regional Young and Senior Water Professionals Conference, Al-maty, pp. 79-93.

Karthe D., Heldt S., Houdret A. and Borchardt D. (2014): IWRM in a country under rapid transition: lessons learnt from the Kharaa River Basin, Mon-golia. Environmental Earth Sciences. doi:10.1007/s12665-014-3435-y

Karthe D., Heldt S., Rost G., Londong J., Ilian J., Heppeler J., Khurelbaatar G., Sullivan C., van Afferden M., Stäudel J., Scharaw B., Westerhoff T., Dietze S., Sigel K., Hofmann J., Watson V. and Borchardt, D. (2015): Modular Concept for Municipal Water Management in the Kharaa River Basin, Mongolia. In: Borchardt, D.; Bogardi, J. & Ibisch, R. (2015): Integrated Water Resources Management: Concept, Research and Implementation. In press.

Karthe D, Sigel K, Scharaw B, Stäudel J, Hufert F and Borchardt D (2012): Towards an integrated concept for monitoring and improvements in water supply, sanitation and hygiene (WASH) in urban Mongolia. Water & Risk 20:1-5.

Ou, Z., Sun, T., Li, P., Yediler, A., Yang, G., Kettrup, A. (1997). A production-scale ecological engineering forest system for the treatment and reutilization of municipal wastewater in the Inner Mongolia. In: Ecological Engineering, 9, pp. 71-88.

Qu-Xing, Z., Qian-Ru, Z., Tie-Heng, S. (2006). Technical Innovation of Land Treatment Systems for Municipal Wastewater in Northeast China. In: Pedosphere, 16(3), pp. 297-303.

Schmidt, G. (2000). Methoden und Techniken der Organisation. Gießen: Verlag Dr. Götz Schmidt.

Sigel, K., Altantuul, K., Basandorj, D. (2012). Household needs and demand for improved water supply and sanitation in peri-urban ger areas: The case of Dar-khan, Mongolia. In: Environmental Earth Sience, 65(5), pp.1561-1566.

USDA (1993). United States Department of Agriculture. National Engineering Handbook., Irrigation Water Requirements, pp. 98-129.

WMO (2013). World Metrological Organization, online available: http://worldweather.wmo .int/119/c01148.htm, date: 12.12.2014.

Westphal, K. (2013): Scenario development of a large-scale willow based wastewater treatment for the village of Khongor in Mongolia with special focus on design and dimensioning. M.Sc. Thesis, Brandenburg University of Technology Cottbus.

Wu S., David A., Liu, L., Dong, R. (2011). Performance of integrated household constructed wetland for domestic wastewater treatment in rural areas. In: Ecological Engineering, 37, pp. 948-954.

VI. Innovative monitoring techniques

Leman-Baikal: Remote Sensing of Lakes Using an Ultralight Plane

Y. Akhtman[†], D. Constantin[†], M. Rehak[†], V. Nouchi[†], M. Tarasov[‡], G. Shinkareva[‡], S. Chalov[‡] B and U. Lemmin[†]

[†]Ecole Polytechnique Fédérale de Lausanne, Department of Environmental Engineering, Lausanne, Switzerland (Email: yosef.akhtman@epfl.ch)
[‡]Lomonosov Moscow State University, Moscow, Russian Federation

Abstract

The paper describes the remote sensing methodology and the initial results obtained as part of the Leman-Baikal project, which constitutes an international Swiss-Russian collaborative research initiative in the field of physical limnology. The three-year framework involves the development and deployment of a novel multispectral and hyperspectral remote sensing platform optimised for the sensing of land and water surfaces from an ultralight aircraft.

Keywords - Remote sensing, hyperspectral imaging, photogrammetry, limnology, resource managements, ultralight plane

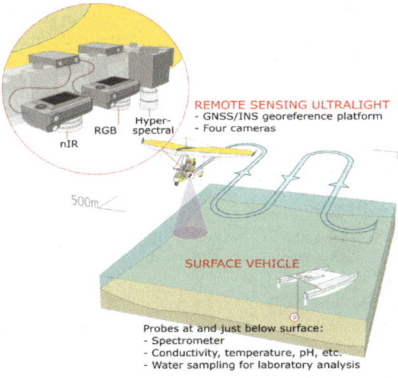

Figure 1. Concurrent airborne and surface data acquisition.

Introduction

Remote sensing technologies provide some of the most effective methods for the exploration and study of the Earth surface [1]. In particular, multispectral and hyperspectral space-bourn and airborne observations are widely used to study different natural and anthropogenic processes [2] including those pertaining to water bodies [3]. The recent technological advances that make remote sensing equipment ever more accessible have brought about a new surge in the interest towards the development of novel and powerful remote sensing methodologies. Of particular interest in this context is the emergence of multiscale analysis, where data from multiples sources: satellites, aircrafts and ground sampling measurements representing different spacial and temporal scales are correlated and jointly processed [4].

The lake Baikal area has a long history of successful application of the various remote sensing methods. For example, Sutyrina used the data collected with the Advanced Very High Resolution Radiometer (AVHRR) to map the ice conditions on the Lake Baikal between 1998 and 2012 [5]. The influences of climatic changes and human activity were further explored by Korytny et al. in [6]. Ivanov et al. studied the Selenga river delta area using Landsat data in aim to detect delta configuration changes from 1701 to 2000 [7].

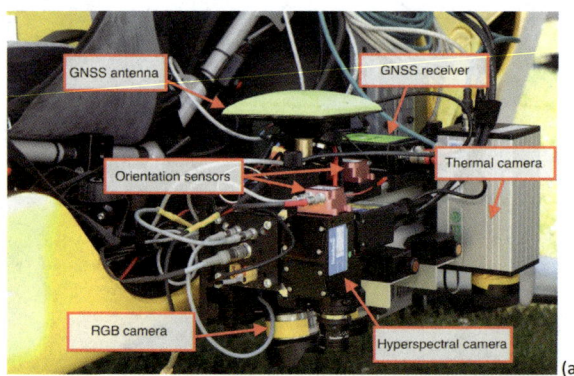

(a)

(b)

Figure 2. Multispectral and hyperspectral remote sensing platform (a) installed on an Air Creation Tanarg 912S ultralight aircraft (b).

Against this background, in this paper we discuss the methodology and the initial results obtained during the first phase of the Leman-Baikal project that took place during the spring and summer months of 2013 on lakes Geneva in Switzerland and Baikal in Russian Federation. The primarily aim of the project is to conduct a comparative study of the functioning of both lakes. The scientific objectives of the project include the analysis of hydrological processes, such as the runoff dynamics of both natural and anthropogenic origin, lake energy balance, and the study of processes pertaining to the land-water and air-water interfaces in lakes.

Methodology

The main principle of the research methodology is constituted by the concurrent acquisition of airborne wide-area and surface point-based data as illustrated in Figure 2. Specifically, we have employed the ultralight aircraft in order to carry an airborne remote sensing platform, and a boat equipped with a range of sensing and water sampling equipment.

As part of the Leman-Baikal project a remote sensing platform was developed to collect multispectral and hyperspectral observations of both land and water surfaces from ultralight aircraft. The platform is comprised of four cameras, auxiliary position and orientation sensors, as well as data recording equipment. Our main instrument is constituted by a Headwall Photonics Micro

Hyperspec VNIR sensor. In addition, the platform includes two high-resolution RGB and near-infrared sensors based on consumer-grade Sony NEX-5R cameras, as well as a thermal infrared sensor based on the DIAS Pyroview 640L Compact camera. The resultant remote sensing platform is portrayed in Figure 2(a). As our airborne carrier we have utilised the Air Creation Tanarg ultralight aircraft depicted in Figure 2(b).

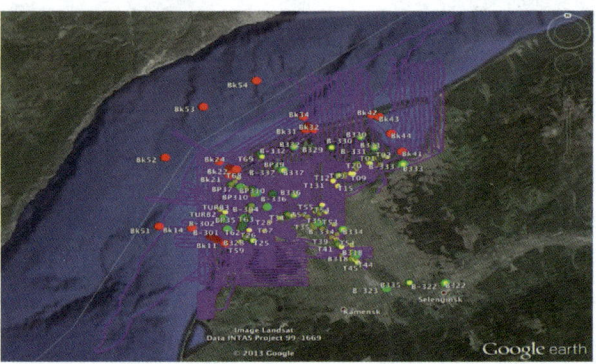

Figure 3. Flight trajectories (purple) and in situ sampling sites (red, yellow and green) for the Selenga delta during the Lake Baikal phase of the Leman-Baikal project, which took place during July of 2013.

The surface-based samples were used to produce a detailed characterisation of the water properties at sampling locations. Additionally, the reflected spectral response of the water surface at each sampling point is registered. The reflectance properties are correlated with the various water characteristics and the spectral response-based indicators for the various chemical and biological water properties are derived. The resultant spectral signature-based indicators are subsequently utilised in order to derive a wide-area maps of water properties using the multispectral and hyperspectral data collected with the use of the airborne remote sensing platform.

In this context, the concurrent airborne and surface based data acquisition methodology exemplified in Figures 1 and 3 is essential for the sake of calibration of the airborne data, as well as the analysis of data quality, accuracy and precision. Correspondingly, ground sampling sites were chosen within the trajectories of the aircraft where the strongest variability of water quality parameters could be observed. We used three radiometers to validate the hyperspectral acquisition from the ULM: the OceanOptics USB 2000+, the waterInsight WISP 3, and the Ramses TriOS system. In addition, water quality parameters were measured with a Seabird CTD19+V2 for chlorophyll-a and turbidity and sub-surface water samples for the Yellow substances. This latter parameters are used as supervision in a neural network which interpret the ULM hyperspectral data in terms of water constituents.

Data acquisition and initial results

During the stage of the system development, as well as during the collection of the initial data, we conducted a series of flights in the area of Lake Geneva in western Switzerland in April and May of 2013. Our initial points of interest included the mouths of the Venoge and the Rhône rivers, which exhibit a particularly rich range of visually observable hydrological phenomena.

(a)   (b)

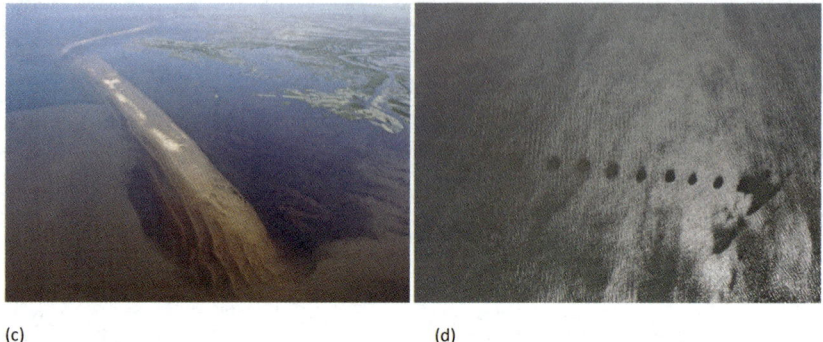

(c)   (d)

Figure 5: Examples of hydrological structures and diverse phenomena observed in the Selenga delta in the course of the field campaign of 2013.

In the consecutive stage of the project, taking place during the months June and July of 2013, we have carried out a comprehensive field campaign in the area of the Selenga river delta in Lake Baikal, which is the largest freshwater reservoir on Earth and is located in the Southern Siberia region of the Russian Federation. The campaign was conducted in close collaboration with the Geography Faculty of the Moscow State University and the Institute of Nature Resource Management in Ulan-Ude. Our airborne observations were complemented by extensive groundwork, which included the collection and analysis of in situ samples, as well as the recording of the corresponding spectral reflectance signatures of the water surface.

A wide range of hydrological structures and diverse phenomena exemplified in Figure 5 have been observed in the course of the field campaign of 2013. The analysis and interpretation of these results is ongoing.

The observations were further resumed on lake Geneva in March 2014 with the aim of recording the seasonal variations in the environmental state of the lake's surface and the associated hydrological processes. Given the large size of the acquired dataset, data processing is still underway. However, preliminary results show a high similarity of the spectra measured from the air and in situ from the lake surface. These encouraging results will soon allow assessment of the heterogeneity of water quality parameters on large portions of the two lakes, and to describe local mixing phenomena at a high spatial and temporal resolution.

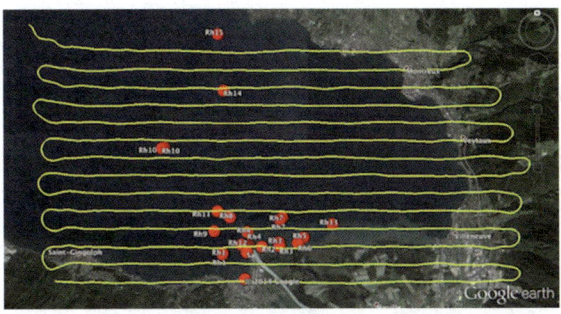

(a)

(b)

Figure 6. Data acquisition campaign of March 7, 2014 over eastern side of Lake Geneva, Switzerland: (a) flight trajectory (yellow) and in situ sampling sites (red); (b) three- dimensional visualisation of the orthorectified and georeferenced hyperspectral data.

Data processing chain

The airborne remote sensing data processing chain is being actively developed for the effective analysis of the material collected in the course of the various phases of the field campaign. The raw data is comprised of multiple data types including multispectral image sequences, hyperspectral line scan sequences, as well as auxiliary navigation data logs. The aim of the data processing methodology is the production of a data management system, which will facilitate access to synchronised, calibrated, as well as time and space referenced multimodal data.

(a)

(b)

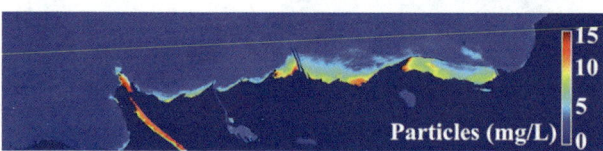

(c)

Figure 6. Visualisations of the eastern side of Lake Geneva including the outflow of river Rhone based on the hyperspec- tral imaging data collected on March 7, 2014: (a) principle component analysis (PCA)-based false-colour; (b) integrated RGB; and (c) neural network-based estimation of the particle concentrations.

For example, major steps that comprise the implemented hyperspectral data processing pipeline include:

- geometric and radiometric corrections of the individual scan lines that compensate for both lens- and sensor- related radiometric and geometric distortions;

- atmospheric correction that accounts for the specific lighting conditions, as well as the effects inflicted by the downwelling and upwelling propagation of light through the atmosphere;
- evaluation of the surface reflectance from the irradiance that takes into account specular effects and mitigate sun glint;
- DTM-based orthorectification and georeferencing.

The performance of the resultant data processing chain is exemplified in Figure 4 that details the flight trajectory, ground sampling sites, as well as the corresponding hyper- spectral data obtained on March 7, 2014 on the eastern side of lake Geneva. It should be noted, however, that not all stages in the detailed pipeline have been completed thus far and the development remains a work in progress.

Furthermore, as part of the Leman-Baikal project we have developed and deployed a dedicated database system and a web-based GIS data management framework, which facilitates an effective and highly structured storage, search, retrieval and visualisation of multi-modal scientific data collected in the course of the field campaign.

Ground truthing resulted in 79 ground control points in Lake Geneva around Venoge and Rhone rivers, and 36 sites in the Selenga region of Lake Baikal. The initial analysis of the obtained hyperspectral data is exemplified in Figure 7 that represents the data recorder on March 7, 2014 in the eastern side of lake Geneva. Specifically, Figure 7 (a) portrays the principle component analysis (PCA)-based visualisation of the hyperspectral data. Figure 7 (c) quantifies the particle distribution corresponding to a typical structure of the outflow of Rhone river in the lake. Notably, the data obtained on March 7 were collected at a time when the photosynthetic activity is still negligible in the lake and therefore our analysis showed no quantifiable chlorophyll distributions. The integrated RGB rendition of the hyperspectral data is provided in Figure 7 (b) for the reference.

Conclusions and future work

In this paper we have presented the preliminary results obtained during the first year of the three year framework of the Leman-Baikal project.

The data acquisition campaigns resulted in the collection of the total of around 7 Terabytes of airborne remote sensing data covering the area in excess of 2000 km², including more than 100 in situ sampling sites. The entire field campaign spanning both Lake Leman and Lake Baikal phases included over 83 hours of flight having an accumulate flight trajectory length in excess of 7,700 km. In particular, the data collected to date is comprised by 580,000 airborne images and nearly 15,000,000 hyperspectral scan lines.

The main focus during the first season of the project was on the development of the remote sensing equipment, as well as the corresponding data acquisition and processing methodologies. Our initial results show a great potential of the developed system. The analysis of the data collected during the 2013 season is ongoing, and the remaining two years of the project offer an outstanding opportunity for further scientific research.

Acknowledgement

Both financial and logistic support of this study by the Foundation pour l'Etude des Eaux du Léman (FEEL), Ferring Pharmaceuticals, the Consulat Honoraire de la Fédération de Russie à Lausanne, as well as the Dr. Paulsen Foundation and the Lake Baikal Protection Fund is gratefully acknowledged. The authors are grateful for the help provided by D. Tuia and D. Ziegler of EPFL, as well as A. Ayurzhanaev, E. Garmaev, and A. Tulokhonov of BINM. Special thanks are due to the team of French and Russian ULM pilots F. Bernard, J. Couttet, A. Barisevsky, A. Sherbakov, V. Vikharev and N. Belyaev, who made this research possible.

References

1. J.B. Campbell and H.W. Randolph, Introduction to remote sensing, CRC Press, 2011.
2. M.E. Schaepman, S.L. Ustin, A.J. Plaza, T.H. Painter, J. Verrelst, and S. Liang, "Earth system science related imaging spectroscopy—an assessment," Remote Sensing of Environment, vol. 113, pp. S123–S137, 2009.

3. S. Koponen, J. Pulliainen, K. Kallio, and M. Hallikainen, "Lake water quality classification with airborne hyperspectral spectrometer and simulated meris data," Remote Sensing of Environment, vol. 79, no. 1, pp. 51–59, 2002.

4. A. Lausch, M. Pause, I. Merbach, S. Zacharias, D. Doktor, M. Volk, and R. Seppelt, "A new multiscale approach for monitoring vegetation using remote sensing-based indicators in laboratory, field, and landscape," Environmental Monitoring and Assessment, vol. 185, no. 2, pp. 1215– 1235, 2013.

5. E. Sutyrina, "Assesment of the transformation of landscapes in the catchment area of the Lake Baikal using satellite remote sensing data," Privolgskij scientific bulletin, vol. 2, no. 30, pp. 195–197, 2014.

6. L.M. Korytny, O.I. Bazhenova, G.N. Martianova, and E.A. Ilyicheva, "The influence of climatic change and human activity on erosion processes in sub-arid watersheds in southern east siberia," Hydrological Processes, vol. 17, no. 16, pp. 3181–3193, 2003.

7. V. Ivanov, V. Korotaev, and I. Labutina, "Morphology and dynamics of Selenga river delta," Moscow University Bulletin. Geography., vol. 5, no. 4, pp. 48 – 54, 2007.

Advantages of Biosensor Water Quality Monitoring

Konrad Siegfried[1], Andreas Koelsch[1], Eva Osterwalder[1], Sonja Hahn-Tomer[1]

[1]Department of Environmental Microbiology, Helmholtz Centre for Environmental Research, Permoserstrasse 15, 04318 Leipzig, Germany

Correspondence: Dr. Konrad Siegfried
email: konrad.siegfried@ufz.de

Abstract

The causes and processes of drinking water contamination with heavy metals, metalloids and other elements have been discussed extensively. In many regions the water monitoring and treatment networks are still not covering the needs of the population. Some reasons are lack of financial resources, non-existing laboratory infrastructure and also missing awareness of authorities as well as population. Another important aspect is the complicated and expensive technology, which is not adapted to circumstances in many of the concerned countries and remote landscapes. The ARSOlux biosensor for arsenic detection in water samples developed by scientists of the Helmholtz Centre for Environmental Research – UFZ, Germany and the University Lausanne, Switzerland is one example of alternative biosensor technologies. It could contribute to the improvement of drinking water monitoring and supply especially in less developed and remote areas. Main advantages are: the detection without use of toxic chemicals, no toxic waste, report of chemical bioavailability, low cost, low material requirement and the simplicity of handling.

1. Introduction

Countless studies have been completed and published about occurrence of arsenic in drinking water in several countries around the world (Ravenscroft et al., 2009). Arsenic in drinking water originates from mining activities and naturally occurring elevated background contents in sediments, which can be released to the ground water under specific environmental circumstances

(Smedley & Kinniburgh 2002). Arsenic is a metalloid, which can negatively affect the health of human beings if consumed in elevated concentrations above the WHO threshold for drinking water of 10 µg/L over an extended period of time. Arsenic contaminated regions in Asia can mainly be found in countries such as Russia, Mongolia, China, India and Bangladesh.

Some severely affected arsenic contaminated areas are found in Inner Mongolia, Shanxi and Xinjiang in China (Rodriguez-Lado et al., 2013), in the Southern Gobi (Olkhanud 2012), the Dornod Gobi (Nriagu et al., 2013), in the North Central part of Mongolia (Pfeiffer et al., 2014) and in several countries of the South- and Southeast Asian Arsenic Belt (SSAAB, Nordstrom 2002). In Russia some arsenic contaminated sites are located in the Ural Mountains (Gelova 1977; Smedley & Kinniburgh 2002), the Northern and Eastern Caucasus (Kortsenshteyn et al., 1973), the Trans-Baikal region (Kuklin & Matafonov 2014), Kovu Aksy in the Tuva Republic (Gaskova et al., 2003), Kamchatka (Ilgen et al., 2011), Chelyabinsk and Kemerovo (Yurkevich et al., 2012). There are manifold origins of arsenic contamination in Russia reaching from natural geothermal and weathering sources to mining and chemical weapon deposits (Henry & Douhovnikoff 2008).

The arsenic issue will serve as an example to demonstrate the advantages of newly developed detection methods based on biosensors containing genetically modified bioreporter bacteria.

Conventional technologies for arsenic detection

Several chemical test kits based on the Gutzeit method are still being used in the field. These cumbersome and rather old fashioned kits require large amounts of material, which in some cases have to be purchased separately. Differences in reagent quality can lead to deviations between measured values and real concentrations observed in a sample. For many of these kits inaccurate results have been observed because readings are based on yellowish to brownish colour scales. Thus, concentration levels are difficult to distinguish by the bare eye (Kabir 2005). The required sample volume is rather high, producing large volumes of toxic waste after finishing the test procedure. Nevertheless, some of these kits (Arsenator®, Wagtech, UK) have been improved in terms of accuracy and use of less toxic reagents. They yield results in good agreement with the reference standard method (ICP-MS) in the laboratory (Safarzadeh-Amiri et al., 2011).

In the laboratory methods such as HG-AAS, ICP-MS, ICP-OES and coupled HPLC with ICP-MS are able to detect total arsenic concentrations, arsenic species and other arsenic compounds with very high accuracy (Mattusch et al., 2000; Hung et al., 2004). These expensive technologies are frequently used in research institutions, commercial and government laboratories in countries where laboratory infrastructures, financial resources and skilled experts are available. However, many laboratories in less fortunate regions in Asia cannot afford the acquisition of high-tech equipment. Moreover, a lack of trained staff can prohibit the utilisation of such sophisticated technology in general.

If appropriate detection and monitoring techniques are not available, no sustainable mitigation of drinking water contamination is possible. Aid agencies, NGOs and governments introduce alternative water treatment methods, which in many cases are not maintained properly due to missing monitoring systems. Promising approaches are tools such as SASMIT, ASTRA and the ARSOlux biosensor (Harms et al., 2006; Stocker et al., 2003). The SASMIT method stands for a locally developed arsenic detection and prevention tool based on a sediment colour scale (Hossain et al., 2014). The decision support software ASTRA enables appropriate choice of water treatment and supply schemes adapted to site specific conditions (Szanto et al., 2014).

2. The ARSOlux biosensor

The ARSOlux biosensor based on genetically modified bioreporter bacteria, can detect total bioavailable arsenic concentrations in ground and surface water samples. The light emission of the bioreporter bacteria correlates with arsenic concentrations in water samples (Siegfried et al., 2012; Pfeiffer et al., 2014).

In order to test a sample with the ARSOlux biosensor for the total arsenic concentration a volume of 1 mL is taken from the water source or sample flask with a syringe. If the water shows high turbidity and red colour (elevated iron concentration), it is recommended to filter the sample with a 0.45 – 0.80 µm cellulose acetate filter. The septum rubber stopper of the biosensor vial is penetrated with the needle of the syringe and the sample is emptied into the vial.

The lyophilisate containing the bioreporter bacteria is re-suspended and the bacteria are revitalised. The biosensor vial is marked with the sample number on the septum stopper or crimp seal. When a certain number of biosensors are filled with samples, the time must be recorded. To avoid time delays a maximum of 20 – 30 sensors should be incubated at once. After the incubation start of the first lot, the next lot can be incubated after 15 minutes and so on. The filled biosensors must be incubated for exactly 2 hours at a temperature of 30°C to restart the metabolism of the bacteria. When the incubation period of 2 hours passed, the biosensor vial is inserted into the measuring channel of a luminometer device to measure the light emission of the bacteria. The measurement itself takes only 10 seconds and the arsenic concentration is displayed on the screen of the measuring device. The luminometer device must be calibrated prior to the measurement of samples. This is done with biosensors previously incubated with arsenic standards of 5, 20, 50 and 200 µg/L.

The measuring range of the ARSOlux biosensor is 5 to 200 µg/L arsenic. The optimum incubation temperature lies at 30°C (between 20°C (68°F) and 37°C (99°F)), pH: 6 – 8, salinity ≤ 0.5% and electric conductivity ≤ 9 mS/cm (TDS ≤ 9 g/L). For best results a mobile or stationary incubator is recommended, which guarantees a constant temperature of 30°C. Arsenic standards used for calibration of the luminometer device should be prepared shortly before the execution of the actual measurements. Unfiltered samples must be analysed immediately after sampling to prevent adverse effects of adsorption and co-precipitation of arsenic. Samples should not be acidified or chlorinated before the measurement. In the special case of very high arsenic concentrations, two biosensor vials are required to analyse one sample since high arsenic concentration of >200 µg/L can inhibit the light response of the bacteria and hence lead to false negative results. In this case, the first vial is filled with the undiluted sample and the second vial with a diluted sample (dilution factor 10). If the result of the diluted sample is higher than the result of the undiluted sample than the result of the diluted sample is multiplied by the dilution factor (10). In such a case the result of the undiluted sample is disregarded. By using the described procedure concentrations up to 2000 µg/L As can be detected with satisfying accuracy (Figure 2). After completing all measurements the used genetically modified bioreporter bacteria inside the biosensor vials must be deactivated by using a disinfectant. Accordingly, the syringe is filled with the disinfectant, the septum stopper is penetrated and one drop of the disinfectant is added to each of the used biosensor vials. The used and deactivated biosensors, syringes, needles and other waste are collected in plastic bags and a container. As a last step, everything is autoclaved.

The performance of ARSOlux has been tested and optimized during several field and laboratory campaigns from 2010 – 2014. Samples were taken from potential user countries with arsenic contaminated regions such as Germany, Serbia, Bangladesh (Siegfried et al., 2012), India (Siegfried et al., 2014), Nepal, Mongolia (Pfeiffer et al., 2014, Figure 1) and Argentina. Technical challenges concerning the performance of the biosensors were observed due to the non-uniform quality of the lyophilisate contained in the bioreporter bacteria. Additionally, the reduction of interactions between the bacteria was leading to improvement of the biosensor signal response. The optimization of these processes had positive effects enhancing the accuracy of the kit. A simplified calibration procedure is going to further improve the overall handling of ARSOlux.

(A)

(B)

Fig. 1. Field sampling and analysis with the ARSOlux biosensor at arsenic contaminated sites in the Ore Mountains in Germany (A) and near a gold mine in the Boroo river valley in North Central Mongolia (B).

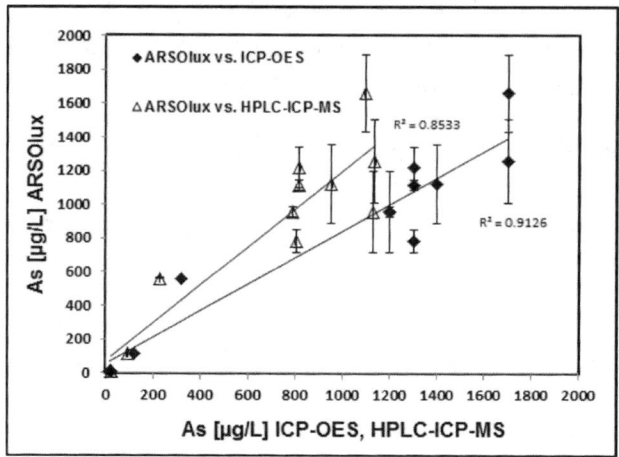

Fig. 2: Cross analysis of samples taken from mining effluents of a former uranium mining site in the Ore Mountains, Saxony, Germany. Samples were analysed with HPLC-ICP-MS, ICP-OES and the ARSOlux biosensor (n=12).

3. Discussion

The ARSOlux biosensor is specifically suitable for usage in screening campaigns and regular monitoring programs. Compared to chemical field kits the biosensor is characterized by higher sensitivity and less material usage. Besides very low volumes of arsenic standards no toxic waste is created during testing procedures. After a short training the handling does not require skills of experts. Moreover, no extra toxic reagents have to be added to the biosensor vial and no toxic reaction products are created. Therefore, the handling is comparably simple, safe and environmentally friendly. It depends on the environmental settings how the biosensor kit is applied best. In countries were laboratories are available in short distance to the testing site, the samples can be transported to the laboratory and be analysed there. In large areas with many water sources and without a nearby lab infrastructure biosensor testing should be done on site. Under such conditions more than 100 samples can be tested in a time period of only one day (Siegfried et al., 2012). The comparison of ARSOlux and a chemical test kit in terms of ability for parallel measurements (screening), need of toxic reagents, material requirements and waste highlights some advantages of the ARSOlux biosensor (Table 1).

The successful commercialization of ARSOlux and similar biosensor systems depends on the approval for import and deployment of the genetically modified bioreporter bacteria included in the biosensor kits. Several requirements have to be fulfilled to successfully introduce such technologies to the free market. The import of genetically modified organisms (GMO) to user countries is regulated by international and national laws and guidelines based on the Cartagena Protocol on biosafety (2000). A risk assessment prepared by the Central Commission on Biologic Safety (ZKBS) of the ARSOlux biosensor for usage in the frame of a research campaign was published at websites of the Biosafety Clearing House and the ZKBS, Germany (ZKBS 2012). The biosensor contains a non-pathogenic *Escherichia coli* DH5α strain. Guidelines and procedures for import permits for research or commercial use vary widely globally. The processing time for applications can take several months or even years.

Independently, an attempt to propose ways for the safe usage of the ARSOlux biosensor under different circumstances will be made. The bioreporter bacteria in the ARSOlux kit belonging to the group of *Escherichia coli* K12 have to be used in laboratories equipped according to WHO biosafety level I (WHO 2004). In Saxony, Germany the biosensor has been used in a mobile lab-van, which was permitted by the Saxon State Ministry of the Environment and Agriculture (SMUL). The lab-van was registered as a mobile genetic engineering facility equipped with manuals, safety sheets and basic protection as well as deactivation and disposal materials such as disinfection flasks and autoclaving containers. A mobile lab could be recommended for regions in India or Bangladesh, where large numbers of wells or other water sources have to be tested in a comparably small area.

	ARSOlux biosensor	Chemical kit
Parallel measurements	>100 samples possible	2 – 6 samples can be analysed at once
Need of toxic reagents	No toxic reagents	Several toxic reagents included (e.g. mercury bromide)
Material requirements	Biosensor vials, Luminometer device, Calibration standards, Syringes & needles, Disinfection solution	Plastic bottles, Mercury bromide test strips, >3 different toxic reagents, Spoons, Canister and bags for collection of toxic wastes, Cleaning materials (brush, water)
Waste	Biosensor vials with 1 mL sample and bioreporter solution (deactivated biomass, LB medium, sugar)	Mercury bromide test stripes, >50 mL sample with toxic reagents, Flushing water

Tab. 8: Comparison of the ARSOlux biosensor and a chemical field test kit for arsenic detection

4. Conclusion

It is obvious that biosensors are a simple and advantageous tool for contaminant detection in drinking water. Based on research results of several measuring campaigns and experimental data from laboratory studies, the optimization process of the ARSOlux biosensor kit for commercial use has nearly finished. Challenges of quality control and handling have been overcome. Currently a novel more appropriate handheld luminometer measuring device is constructed which includes software for data communication and GPS recording. The biosensor kit will be integrated into a multi parameter test kit covering more indicators. New biosensor systems for different settings such as simple semi-quantitative test stripes (lateral flow test), automated online biosensors and sensors for more parameters such as alkanes and BTEX have been developed. It depends on national and international authorities if these promising tools can be made available to a large market and user community in the water sector worldwide. Some of the largest health issues originating from drinking water contamination could easily be mitigated by implementing biosensors.

References

Cartagena Protocol on Biosafety to the Convention on Biological Diversity, Montreal 2000.

Gaskova O.L., Bessonova E.P. and Bortnikova S.B. (2003). Leaching experiments on trace element release from the arsenic-bearing tailings of Khovu–Aksy (Tuva Republic, Russia). Applied Geochemistry 18: 1361–1371.

Gelova G.A. (1977). Hydrogeochemistry of Ore Elements. Nedra, Moscow.

Harms H., Wells M. and van der Meer J.R. (2006). Whole-cell living biosensors – are they ready for environmental application? Applied Microbial Biotechnology 70: 273-280.

Henry L.A. and Douhovnikoff V. (2008). Environmental Issues in Russia. Annual Review of Environment and Resources 33: 437–460.

Hossain M., Bhattacharya P., Frape S.F., Jacks G., Islam M.M., Rahman M.M., von Brömssen M., Hasan M.A. and Ahmed K.M. (2014). Sediment color tool for targeting arsenic-safe aquifers for the installation of shallow drinking water tubewells. Science of the Total Environment 493: 615–625.

Hung D.Q., Nekrassova O. and Compton R.G. (2004) Analytical methods for inorganic arsenic in water: a review. Talanta 64(2): 269–277.

Ilgen A.G., Rychagov S.N., Trainor T.P. (2011) Arsenic speciation and transport associated with the release of spent geothermal fluids in Mutnovsky field (Kamchatka, Russia). Chemical Geology 288: 115–132.

Kabir A. (2005) Rapid Review of Locally Available Arsenic Field Testing Kits; DFID: Bangladesh.

Кортсенсхтайн V.Н., Карасева А.Р. и Алешина АК. (1973) Распределение мышьяка в глубокой грунтовых вод Среднего Каспия артезианского бассейна. Геохимия 4 ,612-617. Publication in Russian language [Kortsenshteyn V.N., Karaseva A.P. and Aleshina A.K. (1973) Distribution of arsenic in deep ground water of the Middle Caspian Artesian Basin. Geokhimiya 4: 612–617.]

Kuklin A.P. and Matafonov P.V. (2014) Background Concentrations of Heavy Metals in Benthos from Transboundary Rivers of the Transbaikalia Region, Russia. Bulletin of Environmental Contamination and Toxicology 92:137–142.

Mattusch J., Wennrich R., Schmidt A.C. and Reisser W. (2000). Determination of arsenic species in water, soils and plants. Fresenius' Journal of Analytical Chemistry 366: 200−203.

Nordstrom, D. K. (2002) Public health - Worldwide occurrences of arsenic in ground water. Science 296 (5576): 2143−2145.

Nriagu J., Johnson J., Samurkas C., Erdenechimeg E., Ochir C. and Chandaga O. (2013). Co-occurrence of high levels of uranium, arsenic, and molybdenum in groundwater of Dornogobi, Mongolia. Global Health Perspective 1(1): 45–54.

Olkhanud P.B. (2012) Survey of Arsenic in Drinking Water in the Southern Gobi region of Mongolia. Master thesis, Johns Hopkins University.

Pfeiffer M., Batbayar G., Hofmann J.; Siegfried K., Karthe D., Hahn-Tomer S. (2014): Investigating arsenic (As) occurrence and sources in ground, surface, waste and drinking water in northern Mongolia. Environmental Earth Sciences. DOI 10.1007/s12665-013-3029-0

Ravenscroft P., Brammer H. and Richards K. (2009). Arsenic Pollution: A Global Synthesis; Wiley-Blackwell: Oxford.

Rodriguez-Lado L., Sun G., Berg M., Zhang Q., Xue H., Zheng Q. and Johnson A. (2013). Groundwater Arsenic contamination throughout China. Science 341: 866–868.

Safarzadeh-Amiri A., Fowlie P., Kazi A.I., Siraj S., Ahmed S. and Akbor A. (2011). Validation of analysis of arsenic in water samples using Wagtech Digital Arsenator. Science of the Total Environment 409: 2662–2667.

Siegfried K., Endes C., Bhuiyan A.F., Kuppardt A., Mattusch J., van der Meer J.R., Chatzinotas A. and Harms H. (2012). Field testing of arsenic in groundwater samples of Bangladesh using a test kit based on lyophilized bioreporter bacteria. Environmental Science and Technology 46: 3281–3287.

Siegfried K., Hahn-Tomer S., Osterwalder E., Pal S., Mazumdar A., Chakrabarti R. (2014). Demonstration of the genetically modified AR-SOlux biosensor in Jyot Sujan, West Bengal, India. In: Litter, M.I., Nicolli, H.B., Meichtry, M., Quici, N., Bundschuh, J., Bhattacharya, P., Naidu, R., (eds.) *One Century of the Discovery of Arsenicosis in Latin America (1914-2014) As2014. Proceedings of the 5th International Congress on Arsenic in the Environment, May 11-16, 2014, Buenos Aires, Argentina.* CRC Press, Boca Raton, FL, p. 867 – 869.

Smedley P. and Kinniburgh D. (2002). A review of the source, behavior and distribution of arsenic in natural waters. Applied Geochemistry 17: 517–568.

Stocker J., Balluch D., Gsell M., Harms H., Feliciano J.S., Daunert S., Malik K.A. and van der Meer J.R. (2003). Development of a set of simple bacterial biosensors for quantitative and rapid field measurements of arsenite and arsenate in potable water. Environmental Science and Technology 37: 4743–4750.

Szanto G.L., van Halem D., Rietveld L.C., Olivero S., Adams A., Roy D.C., Barendse J., Baby K., Hoque M.M. and Dogger J.W. (2014). Decision support for arsenic- and salt-mitigation in Bangladesh: the ASTRA approach. 37th WEDC International Conference, Hanoi, Vietnam.

World Health Organisation. (2004). Laboratory biosafety manual. Third edition. Geneva.

Yurkevich N.V., Saeva O.P., Pal'chik N.A. (2012). Arsenic mobility in two mine tailings drainage systems and its removal from solution by natural geochemical barriers. Applied Geochemistry 27: 2260–2270.

ZKBS. (2012). Statement of the ZKBS on the risk assessment of the ARSOlux test system. Available at http://www.bvl.bund.de/SharedDocs/Down-

loads/06_Gentechnik/ZKBS/02_Allgemeine_Stellungnah-
men_englisch/02_bacteria/ARSOlux.pdf?__blob=publicationFile&v=2.
Accessed 15 December 2014.

The Multi-Species Freshwater Biomonitor: Applications in ecotoxicology and water quality biomonitoring

Almut Gerhardt

Limco International GmbH, Blarerstr. 56, D-78462 Konstanz, Germany

Abstract

Online biomonitors have been developed and implemented in Europe to detect sporadic toxic spills from point polluters, to monitor transboundary rivers at country borders and to safeguard drinking water quality. The Multispecies Freshwater Biomonitor (MFB) is a worldwide unique non-optical biological early warning system to detect toxic spills in all types of water, soil and sediment. The MFB can use all types of animal species, hence represents a cost-effective alternative to the traditional test battery approach. Examples from applications in monitoring of rivers, drinking water and waste water are presented.

Introduction

For several decades the development and implementation of online biological early warning systems has been intensified in Europe caused by (1) several severe toxic spills in rivers, e.g. industrial or wastewater treatment plant effluents, (2) sharpening EU-legislation, e.g. European Waterframework Directive. The development of online in situ methods for toxicological assessment and water quality biomonitoring follows the United nations Agenda 21 (http://www.un.org), highliting the protection and sustainable management of water as a restricted resource (Gerhardt et al. 2006). Whereas standard bioassessment methods based on macroinvertebrate or diatom community structures assess a longterm trend of ecosystem health, online biomonitors can detect pollution waves in real time and allow to take event-related water samples for subsequent chemical profiling in order to trace down the substances in the chemical cocktail which are responsible for the observed biological response in the biomonitor (Gerhardt 1999).Until now a battery of 3 biomonitors (for

Daphnia, fish and mussels) was needed and has been implemented at different stations along German Rivers. However, the MFB can simultaneoulsy use all types of animal species in all media (water, soil, sediment) as bioindicators in one instrument, hence provides a new generation of automated early warning detectors.

Materials & Methods

The MFB is based on the non-optical recording principle, 4-polar impedance conversion (Gerhardt et al. 1994). The test organism is placed individually in a cylindrical flow-through chamber made of acrylic glass. On the opposite inner walls of the chamber two pairs of stainless steel electrode plates are fixed. One pair of plates generated an electrical field by high frequency alternating current, whereas the otherpair of electrodes, at an angle of 90^0 senses quantitatively changes in this electrical field, produced by the movements of the organsism, which itsself functions as a dielectricum (Gerhardt et al. 1994). Different sizes and designs of test chambers can be build to adapt to the ecological needs of the indicator species. For gammarids (up to 1m cm size) we operate chambers of 5 cm length and 2 cm inner diameter. Up to 96 individual channels can be operated simultaneously, i.e. a high number of replication for each indicator species and several different species can be used. The signals are treated for data evaluation and alarm notfication using a specific alarmsoftware: Each signal is analysed by a discrete Fast Fourier Transformation (FFT) to generate a histogram of the percentage occurrence of each signal frequency over the recording period as a so-called behavioural fingerprint. Different types of movements can be distinguished, such as locomotion and ventilation (Fig. 1): locomotion: high amplitudes and low frequencies and ventilation with the gills: low amplitudes and high frequencies. Inactivity provides baselines and mortality is defined as inactivity over a specific time period. Under chemical stress many species show a stepwise stress response: after a 1st avoidance (increased locomotion), the 2nd stress response consists of increased time spent on ventilation, often in combination with elevated ventilation frequency (from 2 .5 Hz normal to 4-6 Hz under stress) (Gerhardt 1999).

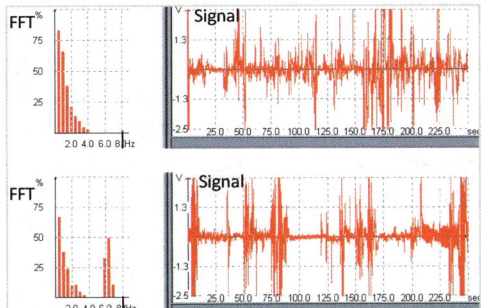

Behavioural signals of *Gamamrus pulex* in the MFB Top: *G. pulex:* normal behaviour (left: FFT, right: raw signal)Bottom: *G. pulex:* stress behaviour (left: FFT, right: raw signal)

Test species

Even though all aquatic invertebrates and vertebrates can be used in the MFB, we often use gammarids due to the following advantages: Gammarus puex/fossarum are sensitive towards pollution (Gerhardt 2012) and they represent bioindicators of class II in the saproby system, i.e. mirroring the „good water quality condition", which is the goal for European streams. Gammarids often are dominant in biomass and abundance at a site and they serve as important prey for fish. Gammarids can feed on leaves and particles, hence feeding in the biomonitor is easy. Their long lifespan allows to monitor also chronic effects of pollution as they can be maintained in the MFB with good oxygen supply for at least 6 weeks. Gammarids have been used in the case studies presented in this paper.

Results/Case studies

The MFB has generated about 25 scientific papers, mostly laboratory studies with specific chemical stressors (e.g. metals, acids, bioban, chlorpyrifos, imidacloprid, diazepam, teramycin), substance mixtures (e.g. metals, oil), complex effluents (e.g. mining effluents, wastewater) and surface water (e.g. Rhine River). Several invertebrate and vertebrate species were tested, Crustacea (11 species), Insecta (8 species), Pisces (6 species), Mollusca (2 species) and worms (7 species), most of them in freshwater, followed b saltwater, soil and sediment.

During the FP6 EU project SWIFT WFD the MFB was placed in a monitoring station along the Rhine River to monitor quantitatively and individually the behaviour and survival of eight adult Gammarus pulex over a period of 6 weeks (Fig. 2). The animals were fed alder leaves. The river water passed through the chambers without pre-filtration, allowing the passage of particles as additional food and particle-bound toxicants to be measured.

Project partners performed regularly chemical analyses of TOC, Flurescence, copper and oil in the water. Changes in the behavioural performance of the gammarids, such as decreasing activity and increased mortality could be linked to elevated concentration levels of the sum parameters and/or specific substances or/and with the response to the Fluotox instrument in most of the alarm cases (Gerhardt et al. 2007).

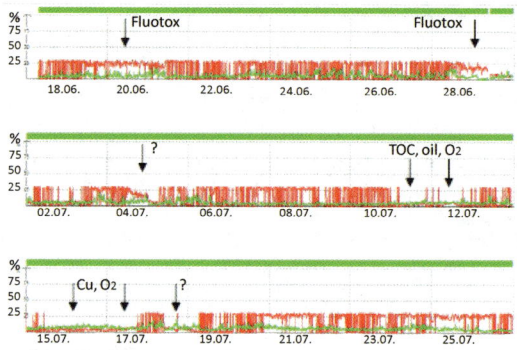

Fig. 2: Biomonitoring of the River Rhine with Gammarus pulex in the Multispecies Freshwater Biomonitor (redrawn from Gerhardt et al. 2007)

Percentage of locomotory activity of G. pulex (mean of 8 organisms)   over time (red line) and ventilation activity (green line). The green bars on top of each graph symbolize the survival of the 8 organisms. Arrows schon changes in the behaviour, which mostly correlate with chemical irregularities.

A second case study shows the application of Gammarus fossarum to monitor the effluent from a WWTP in Switzerland. Eight juvenile gammarids, placed individually in the test chambers and fed by alder leaves could be maintained in the MFB for a period of 2 weeks (Gerhardt et al 2013). Wastewater differs very much in composition and concentrations of the singel substances in the chemical cocktail, therefore effect-based toxicological biomonitoring is the only way to get real-time data of the toxic potential in the water. During an event with high Ammonium concentration, caused by a disturbance of the nitrifying bacteria population in the biological purification step gammarids died (Fig. 3). An event-based water sample revealed high levels of Dithiocarbamates in the water, which induced both the mortality of the gammarids as well as the break-down of the nitrifiers (Gerhardt et al 2014, unpubl. data).

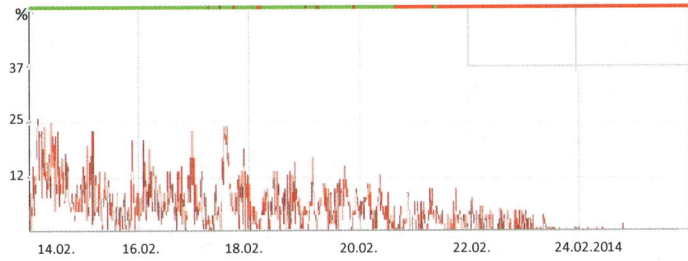

Fig.: 3: Biomonitoring of the effluent from a WWTP (Switzerland): alarm situation (Locomotory activitiy: mean of 8 individuals; red bar: alarm of inactivity phases; green bar: normal behaviour)

Discussion/ Summary and outlook:

Gammarids proved to be excellent bioindicator species in the MFB: robust in handling, sensitive in response to pollution of both inorganic and organic chemical substances. Gammarids occur almost worldwide in different species, both native and invasive species. As their sensitivity towards pollutants is comparable to that of daphnids, they might be applied for monitoring of streams as they are stream inhabitants compared to daphnids as planktonic lake species.

Gammarus lacustris and G. fasciatus also occur in the Selenga river delta resp. Lake Baikal, they appear to be excellent indicator species for water quality monitoring, esp. G. lacustris as its population seems to decline.

Online biomonitoring of water quality in the Selenga river system would be needed to survey the insufficient, incomplete water treatment of waste water in the urban treatment facilities (UNDP 2012). Pollution in the Selenga basin has been attributed mainly to mining activities and ore processing (Cr VI, Hg), pulp industry (chlorinated organic xenobiotics) and urban sources (waste water, coal burnings) (UNDP 2012). As the water from the Selega basin makes about half of the water of Lake Baikal, a good water quality is a necessity to preserve environmental biodiversity (e.g. 1500 endemic species) and human health (drinking water reservoir). Next to biomonitoring at point pollution sources a monitoring station at the country border can provide useful continuous information about the sources of and changes in potential toxic substance loads.

Online biological-toxicological monitoring has been requested by Hofmann et al. (2010) in the highly polluted area around the city Darkhan in the Kharaa basin, just before the confluence of the Kharaa river into the Selenga river.

The Kharaa river basin is affected by several stressors: gold mining in combination with water use and pollution by Hg and other heavy metals; erosion combined with re-location and re-solution of metals from the sediments; eutrophication due to insufficient wastewater treatment (Karthe et al. 2014).

In this area 3 WWTPs, and an ash deposit plant are situated. In this region groundwater, taken for drinking water supply, is polluted by heavy metals, esp. As as well as Chloride and Boron (Hofmann et al. 2013). Metals of special concern originating from gold mining are As and Hg, but also Cd and Pb. As both gold mining and irrigation of the increasing agricultural landuse will further increase a continuous monitoring of the water quality for drinking and irrigation is needed (Hofmann et al. 2010). Some unpolluted headstream areas in the Kharaa river basin situated in the Khentii Mountains (Hofmann et al. 2010, 2013) represent an excellent possibility for a control biomonitoring and source for locally relevant indicator species to be used for online biomonitoring, e.g. in the Buren Tolgoi station.

References

Karthe, D. et al. (2014): Integrating multi-scale data for the assessment of water availability and quality in the Kharaa-Orkhon-Selenga River System. Environment, 65-86.

Gerhardt, A. et al. (1994): Monitoring of behavioral patterns of aquatic organisms with an impedance conversion technique. *Environment International* 20 (2), 209-219.

Gerhardt, A. (1999): Recent trends in online biomonitoring for water quality control. In: Gerhardt, A. (ed): *Biomonitoring of Polluted Water. Reviews on Actual Topics.*, Environmental Research Forum Vol 9, 95-118., TTP Switzerland, 301 pp.

Gerhardt, A. et al. (2006): Field-based effects measures: in situ Online toxicity biomonitoring in water: recent developments. Snvironm. Sci. &Chem. 25 (9), 2263-2271.

Gerhardt, A. et al. (2007): Biomonitoring with Gammarus pulex at the Meuse (NL), Aller (GER) and Rhine (F) rivers with the online Multispecies Freshwater Biomonitor. J Environm. Monitoring 9, 979-985.

Gerhardt, A. (2011): Development of a low-cost ecotoxity test with *Gammarus* spp. for *in* and *ex situ* application. (in: Gerhardt et al. (eds.) see below). In: Gerhardt, A., Bloor, M., Lloyd-Mills C. (eds.: 2011): Gammarus Important Taxon in Freshwater and Marine Changing Environments. International Journal of Zoology: open access http://www.hindawi.com/journals/ijz/osi.html

Gerhardt, A. et al. (2013): Biomonitoring in der kommunalen Abwasserreinigung. Aqua & Gas 7/8, 58-62 (in German only).

Hofmann, J. et. al. (2010): Integrated water resources management in central Asia: nutrient and heavy metal emissions and their relevance for the Kharaa River Basin, Mongolia. WS&T 62.2, 353-363.

Hofmann, J., Rode, M., Theuring, P. (2013): Recent developments in river water quality in a typical Mongolian river basin, the Kharaa case study. IAHS Publ. 363, 2013.

UNDP (2012): Pollution Hotspots in Mongolia: Project „Integrated natural Resource Management in the Baikal Basin Transboundary Ecosystem", final report: VA-2012-78317-010.

EARTH VIEW - GEOGRAPHY AND GEOINFORMATION

Edited by Prof. Dr. Martin Kappas

ISSN 1614-4716

1   *Claudia Sültmann*
     GIS- und Satellitenbildgestützte Landnutzungsklassifikation mit
     Change detection im Westen der Côte d'Ivoire
     ISBN 3-89821-356-0

2   *Katharina Feiden*
     GIS - gestützte Analyse der zeitlichen und räumlichen Verteilung
     der Niederschlagsjahressummen (1961 - 1990) in der Dominikanischen Republik
     Charakteristika und Trends
     ISBN 3-89821-368-4

3   *Nicole Erler*
     GIS- und fernerkundungsgestützte Bewertung von „Natural Hazards" im oberen
     Einzugsgebiet des Rio Yaque del Norte (Dominikanische Republik)
     ISBN 3-89821-409-5

4   *Martin Kappas, Frank Schöggl*
     Bodenerosion in der Dominikanischen Republik
     Eine vergleichende Studie zum Bodenabtrag auf Argrarflächen mit und ohne
     Erosionschutzmassnahmen
     ISBN 3-89821-423-0

5   *Randy Thomsen*
     Change Detection – fernerkundungsgestützte Methoden zur Ableitung des
     Landnutzungswandels in den Tropen (Fallbeispiel Dominikanische Republik)
     ISBN 3-89821-433-8

6   *Sören Steinbach*
     Visualisierung und Quantifizierung von Überschwemmungsbereichen am Mittellauf
     der Elbe
     GIS-gestützte Modellierung von Überschwemmungen
     ISBN 3-89821-530-X

7   *Jobst Augustin*
     Das Seegangsklima der Ostsee zwischen 1958 und 2002 auf Grundlage numerischer
     Daten
     ISBN 3-89821-572-5

ibidem-Verlag / *ibidem* Press
Melchiorstr. 15
70439 Stuttgart
Germany

ibidem@ibidem.eu
ibidem.eu